One Hundred Suns

One Hundred Suns

A Commemoration of the Centenary of the Narayana Gurukula

(1923–2023)

Editor

Swami Vyasa Prasad

DK PRINTWORLD

Publishers of Indian Traditions

Cataloging in Publication Data – DK

[Courtesy: D.K. Agencies (P) Ltd. <docinfo@dkagencies.com>]

One hundred suns : a commemoration of the centenary of the Narayana Gurukula, 1923-2023 / editor, Swami Vyasa Prasad.
pages cm
Includes text in Sanskrit (Sanskrit in roman).
Includes bibliographical references.
ISBN 9788124612835 (HB)

1. Narayana Gurukula. – History. 2. Narayana Gurukula. – Biography. 3. Nataraja Guru, 1895-1973. 4. Nataraja Guru, 1895-1973 – Disciples. 5. Hindus – India – Biography. 6. Hindu philosophy. 7. Spiritual life – Narayana Gurukula. I. Vyasa Prasad, editor. II. Narayana Gurukula, honouree.

LCC BL1226.O54 2024 | DDC 294.544 23

ISBN: 978-81-246-1283-5
First published in 2024

Editor: Swami Vyasa Prasad (b.)

Printed and published by:
D.K. Printworld (P) Ltd.
Regd. Office: Vedaśrī, F-395, Sudarshan Park
(Metro Station: ESI Hospital), New Delhi - 110015
Phones: (011) 4172 2626, 98713 94815
e-mail: indology@dkprintworld.com
Website: www.dkprintworld.com

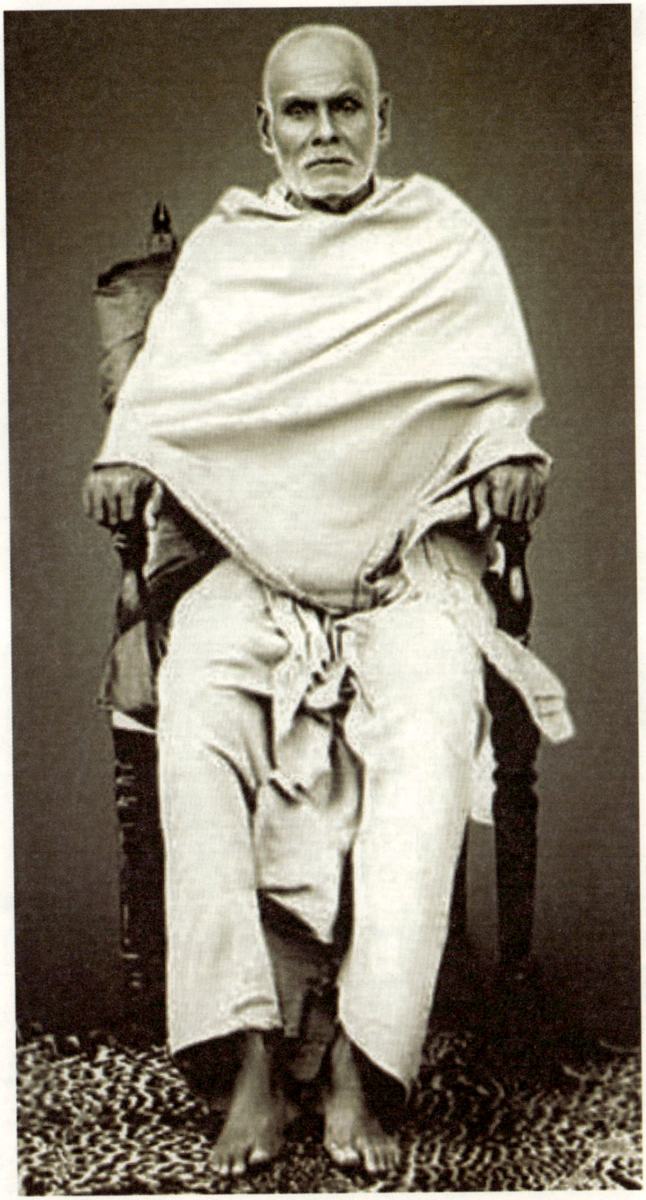

Narayana Guru

Nataraja Guru

Guru Nitya Chaitanya Yati

Guru Muni Narayana Prasad

Swami Thyageeswaran

Swami Tanmaya

Swami Mantra Chaitanya

NARAYANA GURUKULA

ONE FAITH
ONE KIND
ONE GOAL

A
U
M

TO LOVE IS TO LIVE

Preface

FOUNDED by Nataraja Guru in 1923, we celebrate 100 years of history and achievements of the Narayana Gurukula. It is with a sense of gratitude that we present this collection of articles in our Centenary Souvenir. This publication is a tribute to the past, a celebration of the present and a beacon for the future, capturing the essence of our journey over the last century.

Narayana Gurukula, founded 100 years ago, was built on the vision of Narayana Guru and the unwavering commitment and tireless efforts of Nataraja Guru. He laid the foundation upon which we stand today. Over the decades, we have seen many changes, evolving from humble beginnings to adapt and embrace change while steadfastly holding on to our core values and principles.

The articles within this Souvenir are a tapestry of stories, insights and reflections from a diverse array of voices. They chronicle the genesis and maturation of wisdom transforming into peaceful and meaningful lives. The unitive vision of Narayana Guru is the core value and foundation for our understanding. The groundbreaking achievement of Nataraja Guru is his *lingua scientia*, a scientific language for interpretation and integration of the mystical knowledge of the Guru's vision. This innovative stride characterizes the unique perspective of the Narayana Gurukula.

This Souvenir includes contributions from senior disciples from India and the West, and current students. Their narratives not only highlight individual accomplishments but also underscore the collective spirit characteristic of the Gurukula. These pages are filled with personal anecdotes, historical accounts and visionary ideas that offer a glimpse into the vibrant and dynamic community.

The centenary milestone is not just a moment to reflect on our past but also an opportunity to envision and shape our future. The challenges ahead are vast, but so too are the opportunities. With the same resilience, innovation and *tapas* that have brought us this far, we are poised to continue our legacy for the next 100 years.

We extend our deepest gratitude to everyone who has been a part of this project. Your contributions, whether large or small, have been invaluable. This Souvenir is a testament to what we have achieved together.

Finally, in case the reader has not noticed already, ChatGPT assisted in drafting this Preface. Keeping up with the times!

With heartfelt appreciation,

Swami Vyasa Prasad
Narayana Gurukula
Fernhill, Ooty, Tamil Nadu

Contents

Introduction

Swami Vyasa Prasad

A CENTURY is less than a wink of the eye in the flux of eternity. But for humanity a century encompasses an entire lifetime. Nataraja Guru founded the Narayana Gurukula a century ago facing many difficulties establishing Gurukulas in various parts of the world wherever circumstances were conducive. Numerous challenges were overcome by boldly navigating uncharted waters. He called the Gurukulas "neutral islands of unitive understanding" where, by attention to the problems of the individual, all conflicts can be resolved. One can return to what is normal for human beings as nature or neutral reality would have it (*Values* vol. 1.1, p. 25). "Unitive understanding" is a term coined by Nataraja Guru when he was faced with the problem of finding an equivalent expression for what is known as *advaita.* Truth knows no boarders. Wisdom knows no distinction of personal attitudes such as optimist or pessimist, liberal or conservative, orthodox or heterodox, contemplative or active, Eastern or Western. The a priori and a posteriori, analytic and synthetic tendencies, harmonize in a central norm in the wisdom-seeker's way. Ends and means tally unitively and universally when wisdom is fully finalized in terms of a self-happiness which aims at general well-being at the same time. (*Wisdom,* p. 3). In his *Autobiography* (p. 677) Nataraja Guru writes, "All

Gurukulas … depend on the one and only condition above all others which is the dedication of all inmates to the Absolute." The Gurukula is dynamic and open to all who seek in living terms the Wisdom of the Absolute (*brahma-vidyā*) of which the Guru Narayana (1854-1928) is a model and exemplar.

This Centenary Souvenir is a collection of articles written by disciples and devotees who have been touched by the Guru's wisdom in living terms. We begin by Nataraja Guru's article from the very first issue of *Values* published in September 1955. The History of the Gurukula has been written by Guru Muni Narayana Prasad, the head of the Narayana Gurukula. He also explains Structuralism, one of the contributions of Nataraja Guru. This is followed by his Presidential Address given in the presence of former president Shri Ram Nath Kovind and the Governor of Kerala, Janab Arif Mohammad Khan on 20 August 2023. Swami Tyageeshwaran has contributed a poem on Nataraja Guru while Swami Tanmaya has given a detailed account of living with Guru Nitya at the Fernhill Gurukula and his personal journey interacting with several people and institutions. Dr P. Sabu has clarified the Frame of Reference of the Gurukula. Nancy Saumya has shared her pioneering story of founding the Island Gurukula Aranya, the Western Headquarters of the Narayana Gurukula. Scott Teitsworth has written about the Portland Gurukula. Deborah Buchanan has given a delightful account of her experiences with Guru Nitya and Nataraja Guru. Peter Oppenheimer's article is a personal encounter with Guru-wisdom from his youth to mature age. Swamini Brigitte shares her personal transformation under the guidance of Nataraja Guru and the values of the Gurukula. Swami Vyasa Prasad has attempted to present an inclusive view by including contemporary theories. Professor S. Omana has written about Nataraja Guru's magnum opus, *The Integrated Science of the Absolute.* Dr Sugeetha has elaborated on the Element of Wonder in Narayana Guru's Philosophy. Dr Prabhavati Prasannakumar has narrated her experiences with the Gurus and also the application of their teachings in her personal life. Dr Shareena Banu has written a poetical appreciation of Guru's discourse. Swami Mangala Murti has given a historical glimpse into Indian spirituality. Deepanjali's article is a touching and personal account of her experiences with the Gurus and the Gurukula. Gita Gayatri has contributed her understanding of Zen and Narayana Guru's teachings. Peter Moras shares the wisdom he derived from his Hawaiian experience. Bijoyees has a futuristic vision for the Gurukula. Finally appreciations and gratitude for Guru's love and wisdom from a number of people from different parts of the world in twenty-first-century reflections.

The collection of articles show a wide range of experiences and points of view which reflects the open and dynamic nature of the Gurukula. The central theme is contemplative wisdom which is open to revaluation and restatement ensuring that the teachings are relevant to the changing times. The modern world is very tech driven and is transforming at a very rapid rate. Even the machine age is behind us with robots and Artificial Intelligence impacting our lives. Under these circumstances the grounding and steading hand of wisdom has become imperative for the well-being of humanity. There is need for an epistemic journey to discover the true meaning of life. This Souvenir is a small step in that direction.

Narayana Gurukula, Fernhill
December 2023

1

Adoration for an Absolutist Guru

We begin the Souvenir with a short article from the pen of Dr P. Natarajan (later Nataraja Guru) titled "Why Values", which appeared in the first issue of *Values,* released on 11 September 1955. *Values* was a monthly magazine edited and published by John Spiers until June 1974, "Devoted to integrating human understanding unitively, impartially and globally in the interests of the general good". Nataraja Guru's life is best known through his *Autobiography of an Absolutist* in which he narrates his meetings with Narayana Guru at his parent's home, his experiences as a college student, followed by his discipleship which established his bipolar relationship with the Guru. With Guru Narayana's approval, Nataraja Guru established the Narayana Gurukula in 1923 to revalue and restate wisdom which "like a fresh breeze invigorates the static and crystallized consciousness of humanity". It is solely due to the efforts of Nataraja Guru that Narayana Guru's teachings first reached the English-speaking world. The spiritual common sense advocated by Narayana Guru breaks down all barriers that artificially segregate one human being from another.

John Spier's paean is an excellent tribute to Nataraja Guru's, life and teachings:

Through you we call to vision all the Gurus of the past,
From the challenge of the Vedas to Vedānta's counterblast;
From Moheṅjo-Daṛo's yogī *seated underneath the tree,*
To Narayana the Guru from all caste-delusion free.
Through you we know the Buddha, Thiruvalluvar as well,
Great Śaṅkara the expounder, and hundreds more to tell.
Through you we know the Gurus of Ancient Greece and Rome,
Pythagoreans or Stoics, with all we feel at home.
Eckhart, Jalaluddin, Plotinus, Laosi,
Learned Vyāsa of the Gītā, *we know them all through you.*
Though each did write in varied scripts, their teaching is the same;
Like you as Absolutists, they are outside time and name.
Through you we know the secret of their method and their way
Of revaluing the Wisdom grown stagnant in their day.
By your dedicated life for fifty years and more
You have raised on high the teaching of those Gurus gone before.
By your writing and your travels in many a distant land,
You have taught true Guru Wisdom so that all can understand.

Why Values?

P. Natarajan

WISDOM is the highest of human values. The Absolute, whose mystery it is that Wisdom will unravel is not a thing, an entity or even a mere reality. Neither by logical reasoning nor mere philosophical speculation, however systematic, methodical or correct, can that be reached. If, however, the Absolute is not conceived as a human value it would remain a mere abstraction. It could never quench the eternal thirst for knowledge abiding at the core of human nature.

However, when the notion of the Absolute is filled with a living human content, it becomes a matter of absorbing interest above all other interests. These other interests become like small change before the supreme Gold Coin of Absolute Wisdom. It is in this sense that the title of this new magazine should be understood.

The word "values" covers all things that are good, beautiful or true, from the most commonplace to the most sublime. The actual or relational world which is in front of everyone is a feast of varied values. The hunger for values itself has different degrees or kinds. When a clear-sighted man knows what he wants, he is able to select unitive values belonging to every department of life in which he lives and moves. A science of norms in absolute or contemplative values guides each man to select and adopt as his own that unitive value which will give him peace or happiness without coming into conflict with the larger happiness of all life. Thus, man threads his way through a tangled network of values from which he is called upon to select at every given moment.

Through values that are still steeped in the domain of drab necessity, the eye of the contemplative is raised through intelligent appreciation of a unitive scale of values to that highest of states of the spirit in which the Self rests on itself and finds the Absolute Value of all in oneself.

2

History of Gurukulam

Muni Narayana Prasad

WHEN P. Natarajan (later Nataraja Guru) arrived at his home in Ernakulam, he entered the front room and saw Narayana Guru resting on an easy chair with his father, Dr Padmanabhan Palpu, standing behind him. It was an extraordinary scene. Guru stood in front of him, while his father stood behind. In his own words, "Relativism. was receding from my life and Absolutism was coming to the forefront". Natarajan carried an English newspaper announcing his passing the MA and LT degrees together. On hearing the examination result, Guru remarked, "everything is in full agreement".

Narayana Guru, looked like an ordinary country man, wearing a *dhotī* and an upper cloth, but he had already found a place in Natarajan's inner being as a mysterious person with a transparent vision of life. On a previous occasion, Dr Palpu had expressed his wish to dedicate one of his children, or even all of them, for the cause of Narayana Guru. Guru now reminded him of his words. Though there was no response from the father, the son was willing to dedicate himself to fulfil his father's words. A couple of days went by. When Guru was resting in a small hut in the northern region of then Travancore State (possibly in the Parur Taluk), Natarajan approached Guru directly, and after proper salutations, told him that he was dedicating himself to Guru. Guru felt very happy and directed him to join the Advaita Ashram, Aluva. Guru with a smile added, "The Āśrama will have an additional expenditure of just a piece of washing soap every day".

When the present writer was studying in high school between 1951-55, most of the teachers were fresh graduates. After a few years of teaching, they were usually sent to the Teachers Training College so that they could earn their BALT degree. Nataraja Guru already had a MALT (Master of Learning and Teaching) degree which qualified him for a higher post in the field of education. He renounced everything for the cause a great Guru represented. Narayana Guru at that time had no other disciples of such a high stature.

Guru was not always available at the Advaita Āśrama. Those who were taking care of the Āśrama affairs were not mentally prepared to include this new and highly qualified disciple of Guru into their group. His offer to work in the Advaita Ashram School without any renumeration was not approved. Instead, a less qualified person was appointed as a salaried teacher. Moreover, there was a notice displayed at the entrance of the dining hall that those who had not paid boarding fee were not to enter the hall. This put him in more trouble. As the environment was becoming more and more unbearable, he decided to escape from the scene altogether. He made a secret journey to Coonoor in the Nilgiris where Swami Bodhananda, one of the disciples of Guru, was residing. The Swami readily gave him asylum.

While working out the plan of a new model school he recalled a scene which he had come across long ago. After passing his Intermediate examination, Natarajan was once walking past a village in Kerala. He noticed a young man of his own age washing the clothes of his *guru* by the side of a stream. A wealthy man lived in that area and by the side of his house lived a Sanskrit scholar teaching interested students subjects like Vyākaraṇa (grammar), Kāvya (poetry), Nyāya (logic) and the like. The young man was one of his students. His service to his *guru* was the mark of what in India had been known from ancient times as *guru–śiṣya pārasparya* (bipolarity between a *guru* and a disciple). "Why can't this value be given its legitimate place in a modern system of education?" Such was the nature of the challenge before Natarajan at that moment.

During the day he went to a park nearby and meditated on the plans of the school he intended to actualize. By adding the word *kula* to "Narayana Guru", whose disciple he was, and whom he always adored, he found a respectable name for the school. But one aspect he did not think of was that of finance.

With these ideas deeply etched in his mind he went back and presented them to Narayana Guru. The Guru blessed the disciple with three directives:

1. A *gurukula* should be a place where *guru* and disciples live together, forming a family (*kula*).
2. Do not obstruct marriage.
3. The entire world has to become a *gurukula*.

Another directive he had already given to the disciple was, "Go far away, and do great things".

Many experienced people, including Swami Bodhananda, warned Natarajan not to proceed with such a dream-like project.

The Gurukula is Born

The building that was available for the time being to run the *gurukula* was an unused tea factory in Cleveland Estate, near Coonoor, owned by one Rama Swami. He allowed it to be used free of any charge. Thus, the *gurukula* was born in the literal sense. The students were mainly the children of a few people who were interested in such experiments in the field of education. They were twelve in number, hailing mainly from the Trichur area. There were assistant teachers as well. In order to make the two ends meet, Guru and disciples together had to take to the streets for collections by performing group songs.

Before long they were informed that the entire estate, including the factory building, was going to be sold and therefore the Gurukula should be shifted. In the meantime, Nataraja Guru had managed to get four acres of land in the Fernhill area of Ooty with the assistance of some high officials whom he had come into contact with by that time. The students and teachers were shifted to a cooly line which was available nearby. The Head camped in a tent erected in the newly procured land. In 1926, during his visit to the Nilgiris, Narayana Guru reached the hill next to the Gurukula land and looked at it with his blessing eyes, saying "It is nice, it looks like heaven!"

Laying of the Foundation Stone for the Gurukula

The ceremony for laying the foundation stone for the new Gurukula building was held with much *éclat* on 23 June 1926. Gilt-edged invitation cards were sent to important persons of the town. The Maharaja of

Travancore laid the foundation stone. The occasion was his first public speech. Sir C.P. Ramaswami Iyer, the Divan of Travancore, presided over the event.

Before long there came a message from Narayana Guru, who was on his second visit to Ceylon (Sri Lanka), calling Nataraja Guru to settle a dispute among the followers of Guru there. While in Ceylon, Narayana Guru formally gave him a yellow cloth as a mark of recognizing his competence to follow the Guru *paramparā.*

After returning from Ceylon, Nataraja Guru stayed in Varkala so as to have more time to be with Guru. Whenever Guru and disciple met they stayed together for a few days. Guru would teach subtle aspects of wisdom, always continuing from wherever they had paused during an earlier occasion. The disciple would be in the presence of Guru even after the bell announcing lunch went off at the dining hall. The dialogue continued until finally on their last meeting there was a meaningful silence in response to a question by the disciple. It was an indication that the Guru's verbal teachings have come to their term. Such is the way the great Master moulded his disciple, the to-be great Master.

Meanwhile, the Gurukula, started at Fernhill, was virtually following its founder during his years of travel and studies that ensued. The new Gurukula land and the foundation stone lay dormant there, waiting for a new awaking sometime in the future.

The perception of the disciple with regard to education was acceptable to Guru. In order to put it into practice, Guru intended the new school that was being built at Sivagiri to be a model school. But the people who were in the construction committee, which also functioned as the

Managing Committee, appointed another person as the headmaster. Only when the man took long leave to join the law College, was Nataraja Guru appointed as headmaster, during the leave vacancy, at a monthly honorarium of Rs. 45.

It was during these days that Narayana Guru decided to send his dear disciple to the West for higher studies in education. To help him with travel expenses, Guru gave him Rs. 1500 saying, "This money does not belong to any organization or society; we (I) saved it from our own funds earmarked for medicines and *mantras*". This meant that Nataraja Guru would not be accountable to anyone in future on this count. While Narayana Guru, bedridden, was resting in a room at one end of the school building, at the other end of the building, there was a warm farewell gathering of the devotees of Guru. As the young disciple was taking leave of Guru, two bystanders brought two fruits to be given as *prasāda.* Giving them both to the departing disciple, Guru said, "It is good to be so. You will be doubly benefited".

The original plan was to proceed to London from Ceylon. But the will of the Tao decided otherwise, and Nataraja Guru disembarked at Marseille in France, to proceed to Geneva, Switzerland. There he found a job as teacher in a Quaker school in Gland. The curriculum allowed him to freely handle classes in his own way. Subsequently he registered with the Sorbonne for a research programme to gain the title of DLit. His thesis was on "The Personal Factor in the Educative Process". In 1932, he was awarded his DLit. with triple honours. "The personal factor" signifies nothing other than Guru–*śiṣya pārasparya* known in India from ancient days. During this time, he also travelled throughout Europe including England.

On returning to India, it became apparent that he was not acceptable in the *āśrama* founded by Narayana Guru, who had attained *mahāsamādhi*. He went back to his own Gurukula that was waiting for him at Fernhill. He had to live first in a small hut made of tin sheets, with no insulation to protect him from the chilly weather of Ooty. This hut was known as "the ice-box". (This ice-box is retained even now in the Ooty Gurukula). His life there between 1932 and 1949 was one of austerity (*tapas*), penury and intense studies. Meanwhile, with the help of beneficent people, he built the main part of the present main building of the Ooty Gurukula, as if willed by Tao. He was also a member of the Royal Society of Teachers (MRST) by that time. Many who knew him complained that he was wasting his qualifications by not serving the people. He did make several attempts at job-hunting across India. Either over-qualification or intrigues within the inner chambers of palaces prevented him from securing a job.

In the meantime, around 1938, he worked as the Headmaster of Nedunganda High School near Varkala. During this period, he managed to buy two acres of land where the present Gurukula at Varkala stands. Nataraja Guru built a room at the newly bought land and began to travel to Nedunganda by bicycle. Fifty cents were added to this at the time of Guru Nitya, and 10 cents at the time of Muni Narayana Prasad. Another 75 cents of land at a hillock opposite Sivagiri was also bought by Nataraja Guru, where the Brahmavidya Mandir stands today. This was later extended by another 65 cents during the time of Guru Nitya. Returning to Ooty, he continued his austere life and studies at the Fernhill Gurukula. On his fiftieth birthday he took certain definite decisions on what to do next:

1. To write a book about Narayana Guru, meant for those who study seriously.

2. To reshape Narayana Gurukula into an *āśrama* that teaches *brahma-vidyā* and to establish a centre in the West to teach *brahma-vidyā*.
3. To go to America and give lectures on "One Religion".
4. To give perfection to his own knowledge about the East and the West.
5. To collect literature from the United Nations and UNESCO so as to give perfection to his own idea of "One World".
6. These decisions involved adopting a life fully dedicated to *brahma-vidyā* and completely detached from all relativistic pulls. He had no idea as to how all these were going to become actualized.

To America

An invitation was received in 1949 to participate in a "World Conference for Peace Through Religion" to be held in New York, by the Church Peace Union. The invitation was readily accepted, but the means for reaching New York were uncertain. All he had was just Rs. 500 received as his share of the family properties. Unexpectedly Mrs Kumaran Asan visited the Gurukula. On hearing of his predicament, she immediately offered Rs. 2000 as a loan. This solved the problem. She later waived the "loan" and treated it as a gift.

While preparing for his trip to America, John Spiers, his first disciple arrived. Having one disciple, in principle, makes one a Guru. Nataraja Guru thus could entrust the Gurukula with John. John's adopted son Sandy and four more boys became inmates of the Gurukula. In order to meet the daily needs for all of them, Guru says, everything other than wayside robbery was done. When faced with utter needs, what is considered as unjust turns out to be just. Such was the condition of the

Gurukula entrusted with John. Nataraja Guru left for America, travelling by air up to Geneva and then by ship.

After his talk at the World Conference, he visited the Ramakrishna Center in New York and was invited to give a talk. He chose the topic, "How to read the Gita". But as he began speaking, he found it very difficult to develop his subject, as he was covering new ground. He had to end his talk with an embarrassing apology. In the audience, however, was an ex-Norwegian sailor, Harry Jacobson, who became very attracted to Nataraja Guru. Harry was a machine tool maker. He wanted to learn more by becoming his disciple. Guru expressed his desire to start a Gurukula somewhere in America and Harry agreed to have it on his own land in Long Valley in the Schooley's Mountains, a hilly area of New Jersey State, adjacent to New York. Before long Harry had a prefabricated building erected to house the Gurukula. Thus, two wishes were simultaneously fulfilled.

Residing at the new Gurukula, Nataraja Guru commuted daily to the New York City Library. Making use of the library, he dug out the missing links in his studies and integrated them together. In the Gurukula, Harry, his family and some other interested friends, met every evening for discourses and discussions. During this period, Guru gave lectures at various centres in and around New York, including Columbia University. The functions of the Gurukula were thus uninterrupted.

Again to Europe

The Church Peace Union gave a fairly large remuneration making the travel back to India easier. On his return trip, Nataraja Guru decided to halt at Paris where he had acquaintances. On arriving there, an old lady friend offered him a four bedroom flat. In effect, it became a temporary Gurukula. It was during his stay there that he started writing the second part of the book *The Word of Guru*. (The first part comprises articles published in the *Sufi Quarterly* about some incidents in the life of Narayana Guru, written during his first stay at Geneva, while Narayana Guru was still alive.) This second visit to Paris gave the opportunity to renew old acquaintances and also to nurture new ones.

Instead of proceeding to India, Nataraja Guru decided to go back to America. It was on board the ship to America, that he met Garry Davis who had by that time come to the attention of the world as the ex-fighter pilot who renounced his American citizenship and declared himself as a world citizen. Garry had pitched a tent in the United Nations Headquarters in Paris, an international territory. His argument was that this world was already one. The people of the world and their leaders need only admit this fact to ensure a peaceful world. He was deported from Paris to America, and his chance meeting with Nataraja Guru happened on the ship while he was travelling as a deportee. Garry described his stand to Guru, and asked him whether it was wrong.

Guru replied, "If your stand is wrong, my Guru's stand also should be wrong!"

On reaching the Gurukula at Long Valley, New Jersey, the surprise that awaited Guru was a complete set of the *Encyclopaedia Britannica*, an essential help at that particular point of time. Yet another playful help of the Tao!

Garry and his wife visited the Gurukula. From the dialogue Guru had with Garry, there emerged the idea of "geo-dialectics", which means a political science that treated the whole world as one political unit.

Back to India

In 1951 Guru decided to return to India. Instead of flying directly, he decided to stop-over in Europe. The voyage from America to Europe was by a luxury liner. While on the ship he came to know that some of the passengers were on their way to participate in the "World Constituent Assembly of the People of the World" to be held in Geneva very soon. They extended an invitation to Guru to participate in it.

The intention of the Conference was to give shape to a constitution for the people of the world. This idea arose in the mind of one Edgar Gevaert whose house in Belgium was destined to become a Gurukula in future, as though willed by Tao. The Geneva Conference concluded with every participant agreeing for the need of a world government, but nothing materialized.

Back in India

Guru was to board a ship from Marseille to Colombo. The send-off at Geneva was a heartfelt one. The friendships that began in those days were maintained by Guru all through his life. In Colombo, he made friends with many people who had been there with Narayana Guru. Back home, on reaching Varkala railway station, he was received by a huge crowd, including the *saṁnyāsin*s of Sivagiri Mutt, waiting anxiously to see him.

There was a formal reception meeting held at Sashi Theatre in Varkala organized by *saṁnyāsin*s of Sivagiri Mutt and the public of Varkala. A resolution passed in the meeting recognized Nataraja Guru as the successor of Narayana Guru, in the Guru *paramparā*. (In fact, guruhood is not something which is conferred on one by passing a resolution in a meeting of the public. In having a disciple one naturally becomes a *guru*, just as on having a child one becomes a father.)

In the meanwhile, John Spiers by his own effort had procured a piece of land near Kaggalipura in the suburbs of Bangalore city with the aim of starting a new Gurukula centre. He succeeded in starting an institution which later came to be known as the World Centre of Narayana Gurukula.

Swami Mangalananda, then the secretary of Sivagiri Mutt and a gifted orator of unparalleled capability, began to come to Nataraja Guru seeking new points for his forthcoming series of talks. Nataraja Guru said, "This is like a crow flying down and picking up one fish from the basket of a fisherwoman. This will not work. If you are serious, you have to come here and live in the Gurukula carrying out your studies." Without any hesitation the Swami left his position at Sivagiri Mutt and became a disciple of Nataraja Guru.

Another disciple who joined the Gurukula during this period was R. Jayachandran (later known as Nitya Chaitanya Yati).

With Nataraja Guru and the three well-learned disciples at the helm of affairs, Narayana Gurukula in effect became a stronghold of wisdom as envisioned by Narayana Guru.

Meanwhile *The Word of Guru*, the first book from Narayana Gurukula was published.

Gurukula Convention

It was during one of the Sivagiri Pilgrimages that a few seekers arrived at the Gurukula complaining that the Pilgrimage lacked facilities for grasping the wisdom of Narayana Guru. Immediately a discussion group was formed with the visitors as participants. This was the first Gurukula Convention, and it was decided that every year during the last week of December, a Gurukula Convention would be held. This continues to this day, and the most recent Convention was the 72nd one.

The Gurukulas

Nataraja Guru, in those days, was given formal receptions at various centres in Kerala and outside. He received a few offerings of land for starting new Gurukula centres. Thus emerged Gurukulas at Cheruvathur (Kasaragod district) and Engandiyur (Trichur district) in Kerala and at Somanahalli in Karnataka.

Nataraja Guru began to reside permanently at the Ooty Gurukula, and would visit the other Gurukula centres at least once a year, staying in each for a few days. John Spiers moved to Kaggalipura Gurukula and Swami Mangalananda became in-charge of the Varkala Gurukula. Nitya Chaitanya Yati initially lived in Vidyaniketan, the *āśrama* built at his family property, then at Somanahalli Gurukula, then at Premkutir, Bombay, then in New Delhi as the director of the Institute of Psychic and Spiritual Research, and finally came back to Varkala after the demise of Swami Mangalananda. All made it a point to definitely take part in the annual Gurukula Convention at Varkala. After 1968, Nitya Chaitanya Yati began travel abroad teaching in various universities in the West.

Nataraja Guru wrote down a Gurukula Memorandum detailing the mode of functioning of the Narayana Gurukula as a Foundation. Accordingly, the unbroken succession in the Gurukula is to be ensured by the existing head by nominating three successors according to their hierarchy. Further, a Will and Testament are to be made and registered with the Registry. Accordingly, Nataraja Guru appointed John Spiers, Swami Mangalananda and Nitya Chaitanya Yati as his successors.

Meanwhile the monthly magazine *Gurukulam* in Malayalam began to be published from Varkala with Swami Mangalananda as the editor. An English organ of the Gurukula, named *Values*, began to be published from the Kaggalipura Gurukula edited by John Spiers.

Bhagavadgītā Commentary

After reaching India from America, Nataraja Guru began the serious work of writing a commentary on the *Bhagavadgītā.* His main intention was to understand the *Bhagavadgītā* as a Yogaśāstra (science of dialectics) as the colophon claims. The book was published in 1962 by Asia Publishing House, and became the second major publication of the Gurukula.

Singapore Gurukula

In 1956, Nataraja Guru, along with John Spiers and Swami Mangalananda, set out for a journey to Singapore at the invitation of some Gurukula sympathizers there. But at the last moment Swami Mangalananda was denied a visa and could not board the ship to Singapore. Singapore, in those days, was a part of Malaya where combing for communists was intensively going on. Someone hinted to the Malayan Embassy in India that the Swami was a communist, which was the reason behind the denied visa. On reaching Singapore, Nataraja Guru and John Spiers made the facts clear to the authorities and a visa was granted immediately. Swami reached Singapore by flight and joined the team. A new Gurukula centre was opened there. Singapore had a number of Keralites living as migrant workers. The speeches of Swamiji became a craze for them. The three returned to India after touring Malaya also.

I Enter the Gurukula

During all the seven days of the 1955-56 Gurukula Convention, I was given a chance by the Tao to be present at the Gurukula and to attend all the programmes without my planning, in connection with preparing the plan for a temple. It was 1st day of January 1956 that Nataraja Guru accepted ochre cloth and became a *saṁnyāsin.* He did not receive the cloth from any individual. A set of ochre cloth kept ready in front of the statue of Narayana Guru was taken by Nataraja Guru himself.

In 1956 I had yet another opportunity, as a 17-year teenager, to live in the Gurukula for six months. The feeling I had then was, "I belong to the Gurukula". In 1959 Swami Mangalananda allowed me to stay at the Varkala Gurukula and to commute to Quilon daily to work in a government office. After doing so for nine years, I finally relinquished my job and fully became part of the Gurukula.

Guru's Travel to the North

In the early 1960s Nataraja Guru, along with his disciple Nitya Chaitanya Yati, travelled to north India. Guru participated in the Vedanta Conference in Amritsar and a Gita Conference in Bombay. The disciple recorded his experiences during this travel and published them as a book titled *Guruvum Śiṣyanum* (Guru and Disciple). This Malayalam book was the third book of the Gurukula.

The Other Side

Thus while receptions and recognitions were increasing on the one side, attacks and rejections piled up on the other. One man's attack was by using his pen; another's attack was bodily, done by one of Guru's former disciples, who had managed the Varkala Gurukula for a long time. When Guru was camping at the Somanahalli Gurukula, this man hid himself behind the hut in which Guru was sleeping and waited for Guru to come out to relieve himself. On his coming out, the man hit him on the head with a heavy piece of firewood, wounding Guru very seriously. Guru treated it as a punishment for trying to teach wisdom to a person who was not competent (*adhikārī*). Guru had to be hospitalized in order to get the wound healed.

Brahmavidya Mandir

One of Narayana Guru's dreams was to have a Brahmavidya Mandir where all religions are taught. He had even prepared a plan for its construction at the top of the hillock in Sivagiri. But what actually happened was that the original plan was rejected and the Mahasamadhi Mandir, which is meant to be a place of worship, was built over the site where the physical remains of Narayana Guru remain buried. Nataraja Guru, in order to make amends to this error of omission, decided to construct a Brahmavidya Mandir on his own. The foundation stone for it was laid at Guru Narayana Giri on 1 January 1963. A very small semi-permanent room was built on a levelled part of the hill, and I was asked by Guru to live there taking care of the construction works that were to go on. Thereafter, whenever he visited Varkala, Guru insisted on staying in this roomlet ignoring even the non-availability of toilet facilities. Of course, before Guru's second visit, a country-toilet was constructed on the premises. He began to give classes there in the temporary pavilion erected in front of the room on the levelled area. Brahmavidya Mandir thus started functioning there, in its real sense.

Travels

During this period, while making the Ooty Gurukula his permanent residence, Guru travelled to the Gurukulas in Bangalore, Varkala and Cheruvathur spending a few days at each place, giving classes. He planned a journey to Europe too, in preparation for the writing of his magnum opus, *An Integrated Science of the Absolute.*

Guru's 70th birthday was celebrated at the Somanahalli Gurukula in Bangalore in February 1965. A new edition of the *Word of Guru* was also released on that occasion, published this time by Pai and Co., Cochin. After this, Guru along with a few disciples and myself went to Erode to conduct a *guru-pūjā* in a new piece of land that was expected to be granted very soon to start a Gurukula. (This land later turned out unavailable for some unknown reason.)

Thereafter Guru began to prepare for his proposed trip to Europe. The journey was by a Dutch merchant ship S.S. Annenkerk. The ship had twelve special cabins meant for tourists and would take nearly a month to reach its destination, Amsterdam. Swami John Spiers, Swami Mangalananda, Nitya Chaitanya Yati, myself, a large group of disciples and many devotees were present at the Cochin Port to see Guru off. The captain of the ship invited the entire group for formal tea on board the ship.

During this trip the house of Edgar Gevaert, who had organized the World Constituent Assembly in Geneva, became a temporary Gurukula. A tour all over Europe, including England for the second time, was undertaken. There was also an unsuccessful attempt to start a new Gurukula in the southern part of France. Guru returned to India along with Celine Gevaert, the daughter of Edgar Gevaert and Romarin Gracebrook, an English lady.

The Magnum Opus

On returning to Ooty Guru began his serious project of writing *An Integrated Science of the Absolute.* The dictation was regularly taken down and typed out by Fred Haas and Romarin Gracebrook. This first typed script was brought to Varkala. Just after the 1966-67 Gurukula Convention, a team consisting of Nataraja Guru, Nitya Chaitanya Yati and myself sat together at Guru Narayana Giri, going through every page and every word. Three typists were engaged to type the final version without mistakes. This work, with four sitting a day, went on for a month. Three copies of the book were typed as a result of this huge effort. The intention of the book was to introduce a new science of sciences, in which both physics and metaphysics find a common ground for integration. This new science would have an epistemology, methodology and axiology of its own. The heart of this work contains the non-dual vision of Reality of Narayana Guru.

Darśana Mālā

After completing this book, Guru travelled around the world meeting great scientists and discussing with them the possibility of such a science. In most cases he did not get any response. Guru writes in his *Autobiography*:

> ... these views of mind stood isolated because of the new lines of epistemology, methodology and axiology that I had developed with some bold originality derived from Guru wisdom of ancient Upaniṣadic lore.

During this travel he came into contact with the Hippy movement, a new culture of protest against the traditional values of the West. He felt sympathetic towards the movement and a few of them became his disciples. They travelled to India to be with Guru.

Ezhimala Gurukula

Let us now retrace our steps a little in this story. Ezhimala is a hill, 1,000 ft high lying very close to the sea in the Ramanthali island near Payyannur in North Kerala. Between the hill and the sea there is a wide beach. This hill is visible from the premises of the Cheruvathur Gurukula. On knowing of the natural beauty of the area the idea of having a Gurukula there became more and more deep rooted in Guru. While camping with Guru at Cheruvathur in 1960, he said to me, "Let's go to Ezhimala". We started next morning, travelling first by bus up to Payyannur, crossing by ferry and then walking for 5 miles. Guru's pace was very slow, talking to people he met on the way. Somewhere along the way we had some refreshment. By the time we reached the bottom of the hill, it was already late in the evening. When I brought to his notice that it was beginning to get dark, he said, "It doesn't matter, let us climb up". Such was his adventurous nature.

After navigated through bushes in utter darkness for a while, we saw a small lamp burning on the top of the hill, and started walking towards it. We found a middle-aged man sitting alone in a very small thatched hut. He became frightened and wonderstruck on seeing Guru before him in the dark night. He had nothing other than tapioca root to eat. He cooked this immediately and offered it to us along with water as the dinner for that night. Only a gunny bag was available there. It was offered to Guru while we two slept on the unfinished floor. The name of the man living there was Kumaran. Next morning, after visiting the *dargah* of a Sufi, we climbed down the hill. From that day onwards, the desire to have a Gurukula at Ezhimala on the slope towards the sea, became very strong in Guru's mind.

Before long, with the financial help of some lovers of the Gurukula movement, 17 acres of land on the hill sloping towards the sea were purchased – 10 acres at one location and 7 acres at another. A small hut was made there and Guru began living there.

The idea "World Peace Through Unitive Understanding" came to Guru's mind around this time. He planned a "World Conference for Peace Through Unitive Understanding" to be held at the Ezhimala Gurukula, which he named "Gurukula Island Home". As a prelude he formulated a district level conference in 1967, followed by a state level conference in 1968, and an All India Conference in 1969. The conferences lasting for eleven days, began at 11 a.m. on the 11th day of the 11th month, which is Armistice Day, commemorating the end of the First World War. The 1968 Conference was held in Guru's presence. In 1969, Guru was on his world tour and the conference was led by Nitya Chaitanya Yati accompanied by a group of his Australian disciples. For the 1970 World Conference, the chief of the hippy disciples, Curran A. de Bruler (an intellectual), together with Freddy Van Der Borght (an engineer) and Patrick Misson (a rich man) built a two-storied building on an elevated plain of the hillock. On the ground floor there was a conference hall and Guru's room with the dormitory upstairs. A kitchen and dining hall were also built at a lower level. In the meantime, a new Gurukula was started at

Vaithri in the Wynad district of Kerala. The Belgian artist Jean Letchert, together with his artist wife Nicole, lived there and developed it into an art-based Gurukula.

Nataraja Guru and I travelled all through Kerala raising funds for conducting the event. The car we used for travelling was finally ferried to the island along with a barrel of petrol to transport people to and fro for the conference. Later another car from Trichur arrived for transporting participants.

Kumara Swami petitioned the government officials to connect electricity to the venue, which was completed on the eve of the event, by extending electric lines by nearly 3 miles. The road was also made motorable.

The conference had two parts. The daytime programmes in English were conducted in the newly-built hall at the Gurukula. The evening programme meant for local people was held on the beach. It was always followed by entertainment at night. This latter half was managed by Kumara Swami.

When Guru and I were alone after the conference he said to me: "Narayana Guru, towards the end of his life, organized a Parliament of Religions at the Advaita Ashram, Aluva. I, towards the end of my life, have organized this Conference".

During the conference a journalist asked Guru, "Do you think that by organizing such a conference in such a remote area, peace is going to be ensured in the world?" Guru said, "This is a dream of mine. Everyone has the freedom to have dreams. But this world is not so good as to have all my dreams materialized."

Swami Mangalananda Attains Samādhi

A big loss to the Narayana Gurukula movement was the unexpected *samādhi* of Swami Mangalananda, an unparalleled orator, and the second successor to Nataraja Guru. He attained *samādhi* while giving a speech at Piravanthoor near Pathanapuram town in the eastern area of Quilon district. It literally turned out to be the fulfilment of the long cherished dream of Swamiji, "My wish is to die collapsing on the dais while lecturing on the wisdom of Narayana Guru".

The manager of the Gurukula went to Piravanthoor to bring the physical remains of Swamiji to Varkala. I was left at the Gurukula along with a few school boys. Nataraja Guru at that time was resting in a hospital in Kozhikode after undergoing eye surgery. I was alone in the Gurukula to carry out Nataraja Guru's instructions to intern the dead body at one corner of Guru Narayana Giri, where I was living. I made arrangements for digging the pit and proceeding with related matters. The manager returned with the dead body early next morning, but could not be found anywhere thereafter. Perhaps he was sleeping. I had to make all arrangements the next day with the help of the *saṁnyāsin*s of Sivagiri. Swami John Spiers from Bangalore (now Bengaluru) and Nitya Chaitanya Yati from New Delhi arrived the next day.

I took care to follow the directions given in the Narayana Smriti. Accordingly, no function was held thereafter. Narayana Guru directs that the sapling of a tree dear to the demised *saṁnyāsin* is to be planted at that spot. Therefore I planted the sapling of a Mango tree there. Many devotees of Swami Mangalananda were not happy, but I had the full conviction that I was not wrong. I never showed any disrespect for Swamiji.

After three months of Swami Mangalananda attaining *samādhi*, the manager of the Gurukula at Varkala absconded for reason known only to him. Once again Swami Nitya reached Varkala. He handed me the bunch of keys of the Gurukula and said, "This is Prasad's *karma*". Thus, unexpectedly, I became the man in charge of the Varkala Gurukula which was the headquarters of the Gurukula movement. At that time, I was working in a government office in Quilon. Every morning I had to leave the Gurukula and return only by 6 o'clock in the evening. I found holding both responsibilities practically impossible, and so I gave up the government job to concentrate on Gurukula affairs and my studies.

World Parliament of Religions

A World Parliament of Religions was held at Sasthamkotta in Kerala in 1971. It was held in coordination with the Akhila Bharatha Ayyappa Seva Samgham, an organization meant for serving the devotees of Ayyappa, the presiding deity of Sabarimala Temple. This body was then headed by M.K.K. Nair, a well-known administrator, and C.N. Sreekandan Nair, a well-known playwright. The Parliament of Religions was conducted at the behest of these two people. The venue, suggested by Nataraja Guru, was a place much loved by Chattampi Swami, another spiritual Master who was like an elder brother to Narayana Guru.

The ten-day programme was given shape by Nataraja Guru, keeping in mind the structure of the ten chapters of *Darśanamālā* of Narayana Guru. The event was inaugurated by Nataraja Guru, and the entire deliberations were regulated by Nitya Chaitanya Yati. My role was to present two papers, to be the rapporteur of the entire event and help in proofreading all the papers, which were already printed in advance, and also to help with office work.

An area protruding into the large fresh water lake in Sasthamkotta was converted into a small temporary township, with buildings made of thatched coconut leaves. It consisted of an auditorium, a prayer hall, special dining hall for delegates, an ordinary dining hall and kitchen for visitors, and a number of rooms with attached baths with running water. All the government guest houses in the Quilon district were booked to accommodate the delegates. Delegates were seated in a gallery having a table with microphones for the speakers. Separate galleries were there for visitors. In short, it was a big event for which the organizers could congratulate themselves. Another Peace Conference was held at the Gurukula Island Home in November 1971 almost of the same scale as the 1970 Conference. The special guests this time were the Maharaja of Porbandar, General K.M. Cariappa, the first army chief of Free India, and Garry Davis.

Meanwhile, by late 1970, four new Gurukula centres were opened at Tholpetty and Peria in Wayanad and at Alattur and Erimayur in Palghat district. Of these, the Peria centre later developed into a world-renowned botanical sanctuary.

Saundaryalaharī

Nataraja Guru was relaxing after completion of his magnum opus, *An Integrated Science of the Absolute.* By chance he came upon the composition *Saundaryalaharī* of Śaṅkara. He, noticing the uniqueness of the work, examined all the available translations and commentaries. He found that the structural linguistic device the *ācārya* adopted in this work had not attracted the attention of commentators, and thus the true spirit of the work had eluded them. This work became the subject of discussion with everybody he met during his 1969 European tour. He later wrote his own commentary on this unusual work.

Swami John Spiers Leaves Gurukula

Swami John Spiers was the first disciple of Nataraja Guru, as well as his nominated successor. He sent a letter to Nataraja Guru and Swami Nitya informing them that he was leaving the Gurukula movement. A copy was sent to me also as the person incharge of the headquarters. It was quite unexpected. The reason he stated in the statement published in *Values* was that one need not remain a disciple all through his life. *Values* ceased to be the organ of the Gurukula movement, and the Narayana Gurukula at Kaggalipuram was renamed Narayana Guru Ashram. Nataraja Guru, thus found himself in a helpless condition of having only one successor in the place of three. It was in such a forlorn state that Guru one day declared that I should be his second successor. A will to this effect was registered the very same day. This came as a big shock to me. Even though I had been formally initiated as a *brahmacārī* in 1969, how could a person of no maturity, of no proper education and no knowledge, be the successor of Swami Nitya Chaitanya Yati who was at that time a visiting professor in some well-known universities in America. I was helpless and took it as a new challenge in life.

For some reason, Swami Nitya was not accepted as the successor of Nataraja Guru by the hippy disciples of Guru. Consequently I also became unacceptable. They formed a separate group within the Gurukula and chose one among them as Nataraja Guru's successor.

In June 1972, while in Bangalore, Nataraja Guru had a stroke which totally paralysed his left side. I reached there immediately and Swami

Nitya, who then was in Australia, flew in before long. Guru was brought to Trivandrum Medical College for treatment, and thereafter to Varkala Gurukula. It became apparent that Guru would not regain his normal health. Swami Nitya left Varkala for his travel overseas giving me all the instructions in writing on what was to be done, should the inevitable happen. When Guru was in an extremely weak condition Curran de Bruler and Freddy came with a Pontiac car to take him all the way to Ezhimala. The doctor taking care of Guru did not give permission to travel and they returned disappointed. Nataraja Guru attained *samādhi* on 19 March 1973. His physical remains were interred at the central point of Guru Narayana Giri around which the proposed Brahmavidya Mandir was to be built. A *yati-pūjā* (a ritual in honour of *saṁnyāsins*) was held on the sixteenth day and *maṇḍala-pūjā* (worship followed by public feeding) was held on the forty-first day.

Guru Nitya's Time

A general council of the Gurukula disciples was held thereafter as directed by Guru Nitya at Varkala in an attempt to bring unity among the disciples of Nataraja Guru. There was no sign of the hippy group accepting Guru Nitya as the successor of Nataraja Guru. On my part, I could not take a stand that goes against the will of Guru. Thus the Gurukula disciples, at least for the time being, became divided into two groups. The Gurukula Island Home, the Gurukulas in Wayanad and one in Bangalore came to be under the control of hippy disciples because they had all been kept under the charge of one hippy or another by Nataraja Guru himself. The other *gurukula*s remained under Guru Nitya's control.

Before long, the land of the Gurukula Island Home was taken over by Government of India for establishing a Naval Academy. Out the compensation given by the government, bare land value was accepted by Guru Nitya, and the amount corresponding to the expenses incurred for all the developmental works was given to Freddy who was the resident there. Freddy moved to Bangalore and started the Nataraja Gurukula. The other hippies who were in charge of various *gurukula*s, left India one by one.

The Gurukulas in America

The Gurukula started in New Jersey during Nataraja Guru's time was on the land of Harry Jacobson. It continued to function for a few years during Guru Nitya's time also. But after Harry, his children were not interested in continuing at Gurukula and so it became non-functional.

From 1968 to 1985, Guru Nitya was mostly in Australia and America teaching as visiting professor in various universities. During this time *gurukula*s at Bainbridge Island and Portland were started. Their stories have been written in detail by Scott Tietsworth and Nancy Yielding.

On 19 March 1973, when Guru Nitya took over as Guru and Head of Narayana Gurukula, he was away in England. Where a Guru resides is not a matter of any importance; what matters is that he imparts wisdom to seekers who approach him. Nataraja Guru's life was very much of the same nature. By the end of the year Guru Nitya arrived at Trivandrum airport. Grand receptions were arranged by the *saṁnyāsī*s of Sivagiri Mutt and those who loved Narayana Guru, along the entire way till he reached Varkala.

A piece of land had been donated to the Gurukula during Nataraja Guru's time in Ochira. The actual functioning of the Gurukula there started only during the time of Guru Nitya. As if to compensate for the loss of the Gurukula Island Home at Ezhimala, a new Gurukula centre at Kanakamala near Thalassery was established in 1979. It is a spot of scenic beauty from where one can see the sun rising from the Sahya Mountain in the east and setting in the Arabian Sea in the West.

During the same period, a few local people interested in the teachings of Narayana Guru formed an association call Guru Smarana Samiti at Edappally in Kochi. A lady much devoted to Narayana Guru donated ten cents of land in that city area. A small Gurukula began to function there soon after. Similarly a piece of land was bought by a few people of Tripunithura town within the Greater Cochin area and a new Gurukula was started there also.

Once Guru Nitya, while climbing the Malayattoor Hill along with some followers, noticed the natural beauty of the locality. Manappattu Chira, an extensive water reservoir, is at the bottom of the hill. He envisioned that a piece of land on the side of the reservoir would be ideal for a Gurukula meant for peaceful meditation. The people who were with him along with some others shared the cost of 20 cents of land there. A semi-permanent building was raised within a week and a three-day long convention was held there marking the formal opening of the Gurukula centre. After nearly forty years the semi-permanent building was replaced with a permanent one.

As a memorial to Swami Mangalananda, his disciple Kumara Swami established an *āśrama* called Mangalabharathy on the banks of Periyar, at Thottuva in Ernakulam district. After Kumara Swami's demise, this *āśrama* was annexed to Narayana Gurukula.

Vidyaniketan, an *āśrama* built by Guru Nitya Chaitanya Yati in the share of his family properties at Murinjakal, konni, also became part of the Gurukula movement.

Brahmavidya Mandir

The foundation stone of the Brahmavidya Mandir was laid in 1963, but only levelling the top of the hill was done when Nataraja Guru attained *samādhi* in 1973. The construction of the building designed by architect A.K. Jayachandran, began in 1975. The ground floor and library were inaugurated in 1980 by Uthradam Thirunal Marthanda Varma, the first successor to the Maharaja of Travancore. Work on the first storey that houses the auditorium was started in 1982. All the inmates of all the *gurukula*s travelled all across Kerala collecting donations from 50 paise and above. As this work was going on, 65 cents of land adjacent to Guru Narayana Giri was bought and annexed. Though the building is fully functional, its finishing work is not yet over. Narayana Guru's principle regarding such God-related works is that it should go on as a *tiruppani* – unending, bit by bit.

The original English name of Brahmavidya Mandir given by Nataraja Guru was "Institute of the Science of the Absolute". It continues to be the same.

Publications

The Gurukula publications at the time of Nataraja Guru were *The Word of Guru*, *Bhagavadgītā* (commentary), *Guruvum Śiṣyavrum* (Malayalam) and *Ātmopadeśa Śatakam* (meaning and elucidation) by Nitya Chaitanya Yati in Malayalam alone. A small printing unit was bought by the Gurukula in

1974, making it easier to publish books one by one. The articles of Nataraja Guru published in *Values* were collected and published along with their Malayalam translations. Nataraja Guru's works and their Malayalam translations, Guru Nitya's writings in English and Malayalam, and my own attempt to write in both these languages and the writing of some like-minded people were published. At present, there are around 350 titles in the Catalogue of Gurukula Publications when the Gurukula celebrates its Centenary Year.

Guru Nitya concluded his assignments in foreign universities, and settled down at the Fernhill Gurukula in 1985. Thereafter he began to publish many articles of psychological import in many periodicals in Kerala, and enchanted by them many young men and women who were having mental problems got attracted to the Fernhill Gurukula. The Gurukula became an asylum for many of them. As a result, a visitor would see twenty to twenty-five inmates in the Gurukula during those days. Running a big kitchen to feed all of them was yet another problem. This phenomenon continued till Guru Nitya attained *samādhi* in 1999. The most important work he did during this period was his commentary on the *Bṛhadāraṇyaka Upaniṣad* both in English and Malayalam. We had together agreed to write and publish Gurukula's own commentaries on all the ten major Upaniṣads. The *Bṛhadāraṇyaka* was done by Guru Nitya and the remaining nine by me in both Malayalam and English. Guru Nitya up until the last moment of his life continued to radiate the light of wisdom directly to those who were around him, and indirectly through his writings. He even made use of his final illness as a mean to do the same. Nataraja Guru's dreams of the Brahmavidya Mandir was almost completed at the time of Guru Nitya.

The Time of Muni Naraya Prasad

Smt. P. Indira, a disciple of Guru Nitya, was initiated as a *saṁnyāsin* by Guru himself. Her house in Palakkad town was functioning as a Gurukula Centre as well. As she felt that it would not be possible to continue after her lifetime, she bought another 10 cent of land from a Gurukula friend Shri Sivarama Krishnan at Kottekkad, outside the town, and constructed a building suitable for the Gurukula. Later the remaining part of the land was also donated by Sivarama Krishnan to the Gurukula. The centre now functions as a women's *āśrama*.

An important event worth mentioning during this period was the three-day conference held at the Indian Institute of Advanced Study, Shimla, in 2011 on the philosophy of Narayana Guru. Yet another event of the same nature was a two-day seminar on Narayana Guru's philosophy held at the Brahmavidya Mandir, Varkala in 2017, sponsored by the Indian Council of Philosophical Research, New Delhi. Most of the resource persons were from among the researchers of the Gurukula.

Two book releases during this period are unforgettable. One was the releasing of my *Chāndogya Upaniṣad* commentary in Malayalam and English simultaneously at the Kanakakkunnu Palace, Trivandrum. The books were released by Governor of Kerala Shri R.L. Bhatia. The next release was my commentaries on nine Upaniṣads held at V.J.T. Hall, Trivandrum, by Justice P. Sadasivam, the then Governor of Kerala in 2018. Another event was the publishing of my commentaries on the complete works of Narayana Guru by D.C. Books, Kottayam in three volumes. My English translation of the complete works of Narayana Guru was published by the National Book Trust of India, a Government of India undertaking.

Awards

My translation into Malayalam of Nataraja Guru's commentary on *Saundaryalaharī* won the Kerala Sahitya Akademi award for the best translation of the year 2015. Similarly, my autobiography *Ātmayānam* got Kerala Sahitya Akademy's award for the best autobiography of the year 2018. Among other awards, a noteworthy one is the Vedantaratna award given to me by the Advaita Ashram, Kolathur, Kozhikode. In 2024 I was awarded the "Padma Shri" by the government of India for my contribution in the fields of education and literature.

Śatābhiṣeka

My Satābhiṣeka (the chance of having seen 1,000 full moons during one's lifetime) was celebrated in 2022 on a grand scale. The main event was organizing about 100 discussion groups based on my works. Two souvenirs, one in English and one in Malayalam, were also released. An English documentary film entitled *Guru Muni: Being and Becoming*, on my life and philosophy was also made by Gurukula friends in London. Its premier was held at the Bharath Bhavan Theatre, Trivandrum on 8 May 2022. It was released in London also. The film is available on YouTube: https://narayana-gurukula.org/documentary.html

After the time of Muni Narayana Prasad, Narayana Gurukula is to be headed by Swami Tyageeswaran, Swami Tanmaya and Swami Mantra Chaitanya in this order.

3

Structuralism as I See It

Muni Narayana Prasad

ONE of the difficulties the writings of Nataraja Guru presents the reader is his use of pictorial language or structuralism. Let us try and understand this method.

Everything in this world has a structure of its own, some definite, some amorphous and some a combination of both. On encountering these structures, we want to know them. We need true knowledge that will give a true image of the structure of the object concerned. Every object is in a state of change because nothing in this world remains changeless. Hence, the knowledge of an ever-changing object will also be an ever-changing one.

The world as a whole is in a constant state of flux. Yet, it has a structural perfection of its own, though inconceivable to the human mind. Therefore, the true knowledge of the world, if at all it is possible, should also be a changing one while having an innate structural perfection of its own. This means that knowledge should be a moving image of the moving world. It can also be referred to as "the moving image of eternity".

All this pertains to the apparent aspect of reality. Let us now look at the interior of the reality, the realm of the changeless.

Let us take the case of a gold chain. The apparent aspect is the chain, good enough to wear around one's neck as an ornament. The real content in it is gold. The chain-form is subject to change; it can be melted and reshaped into another ornament at one's will. But the gold substance in it can never be changed. It always remains gold, however, it may be shaped. It means that the chain-aspect is fleeting by nature, but the gold-aspect is eternal. Without the eternally existing gold, no ornament of a fleeting nature can be made; likewise, without having some fleeting form, the eternal gold-substance cannot exist. Thus, there is an inseparable oneness of the eternal substance and the fleeting forms.

It is easy to have a knowledge of the fleeting apparent form as well as a notion of an eternal substance. But understanding the togetherness of the two is a contradiction for the reason that the eternal and the transient remaining one is not possible logically. But the two remaining inseparably one is a fact in the present moment. Is there any means of giving expression to this oneness?

With the help of a graph, using the intersection of two axes, one horizontal (X-axis), and the other vertical (Y-axis), we can locate a point, or a bit of information, which is neither vertical nor horizontal. Once we locate the point or information we discard the graph we used. Similarly, in geography we use latitudes and longitudes to locate places or even points in mid-ocean. But no such lines are seen on the surface of the actual earth. Nataraja Guru points out that the same method can be used to know the oneness of the eternal substance and the fleeting apparent forms. This oneness can never be thought of as eternal nor as transient.

Suppose the transient appearance is represented by the horizontal (X) axis, and the eternal substance is represented by the vertical (Y) axis. It is easy to understand the oneness of the two at the point of intersection of the two axes. Though not verbally described, the oneness concerned is now graphically understood.

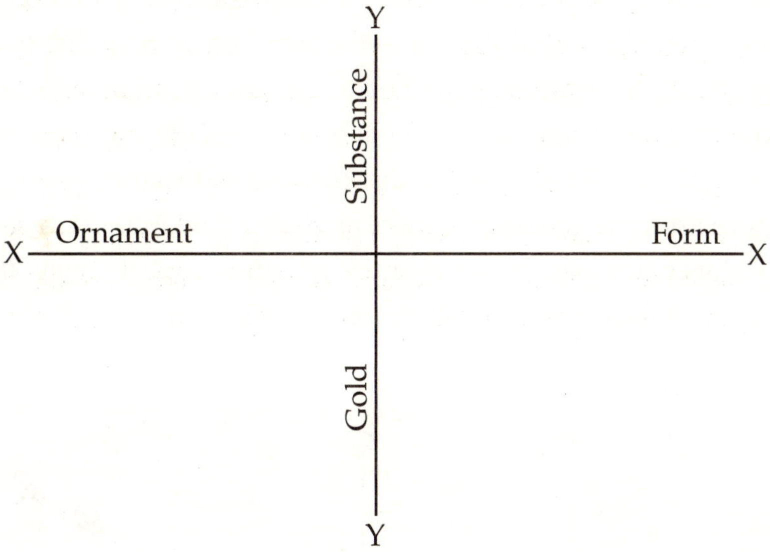

Let us now think of our own existence, of a person wearing this chain. I am a living and constantly changing entity, with an apparent aspect and an invisible and unknowable substance aspect. (Gold as a substance in the being of all ornaments is also invisible for the reason that without a transient apparent form, gold as such is invisible.) The apparent aspect in my being is termed *kṣetra* (field) in the *Bhagavadgītā* and it includes everything physical and mental in me, ranging from the material body to the life principle (*cetana*) (see *Gītā* XIII.5-6). There is an invisible and unknowable substance aspect that assumes all these apparent aspects also in my being, termed *kṣetrajña* (knower of the field) in the *Gītā*. Without this substance, the apparent aspects cannot have being and cannot function; without the apparent aspects, the substance aspect cannot have being either.

There is an inseparable oneness of these two aspects. This non-duality (*advaita*) cannot be verbally expressed other than by referring to the two aspects. Again the graphical method comes to our aid in expressing the non-duality. The horizontal axis (*X*) stands for the *kṣetra* (apparent aspect), and vertical axis (*Y*) stands for the *kṣetrajña* (substance aspect). (This substance aspect is generally known as *ātmā* in Vedānta in the sense it is the substance that pervades the entire apparent aspect (*āpnoti iti ātmā*). The apparent aspect is called *śarīra* (body). Then the non-duality of the

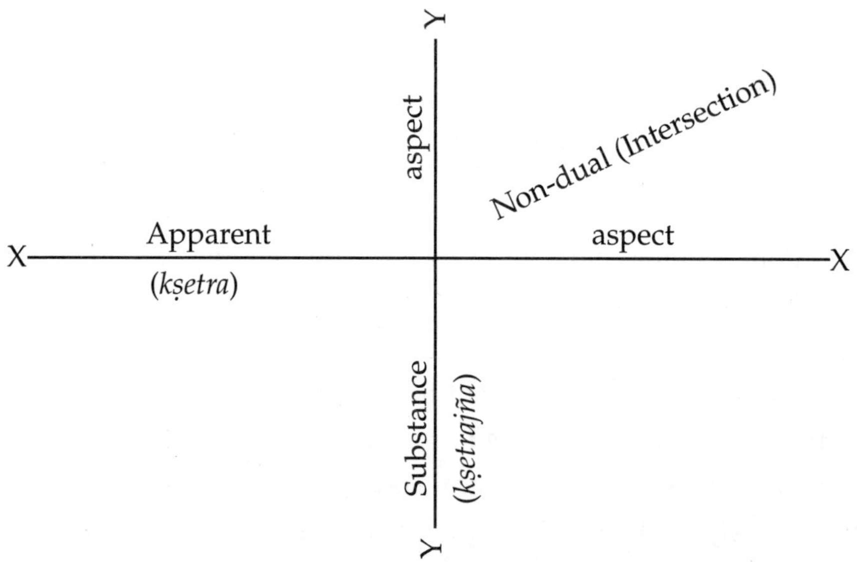

two is graphically marked by the point of intersection of the two which is neither horizontal nor vertical, neither *śarīra* nor *ātmā*.

On the other hand, my being is not like the gold chain, an inert entity. I am a living entity, a creatively existing entity. My own being is known to me, the creative nature of my being is also quite evident to me. But expressing it verbally to someone else is not possible. Here again, pictorial language becomes useful. Creativity generally takes place at the apparent level. There are many ways in which this creativity can become formed. Only three possibilities are shown in the sketch below.

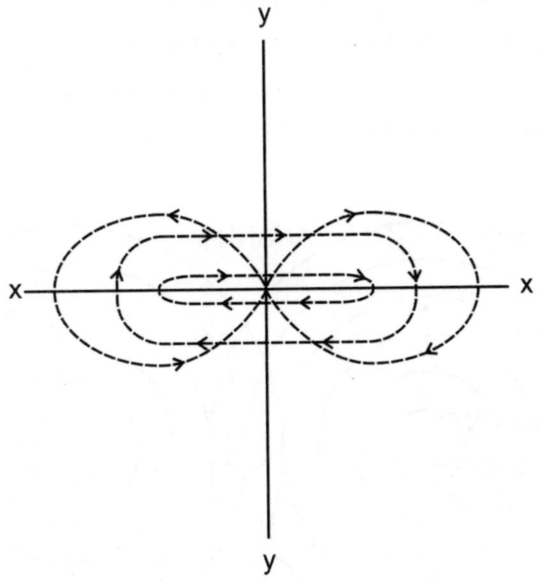

I know myself. I am aware of the creativity in me as well. That means, I perceive myself as a creatively existing being. Every knowledge involves a subject and an object. In the case of I knowing myself, the subject becomes the object and the object becomes the subject. Every knowledge involves two aspects. One is the vertically existing knowing consciousness and the other is the subject and the object meeting horizontally in the knowing consciousness.

In the case of I knowing myself, instead of the subject and the object coming from either side and meeting at the neutral point, both remain at the neutral point, one becoming the other. Here, pictorial language also seemingly fails.

What is the real substance in my being? It is pure and unconditioned consciousness. Narayana Guru calls it *aṟivu.* Some Vedānta texts like the *Jñāna-Vāsiṣṭha* call it *cid-vastu,* meaning Consciousness-Substance. It is defined as *sat–cit–ānanda.* As the basically existing Reality, it is *sat;* as consciousness in essential content, it is *cit,* consciousness itself; as

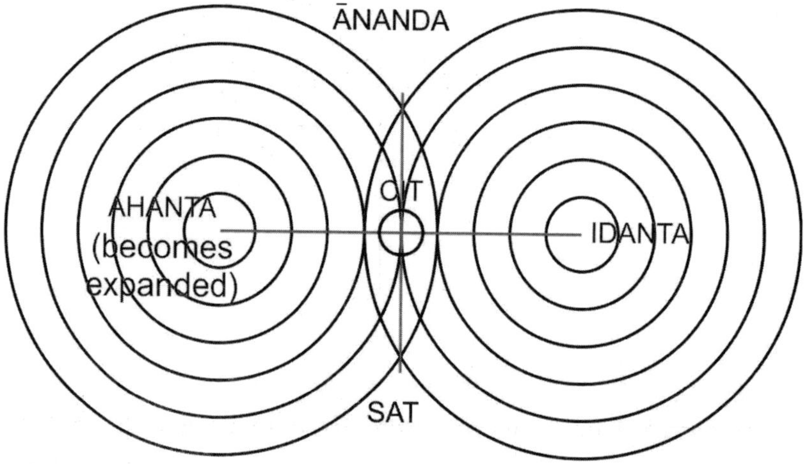

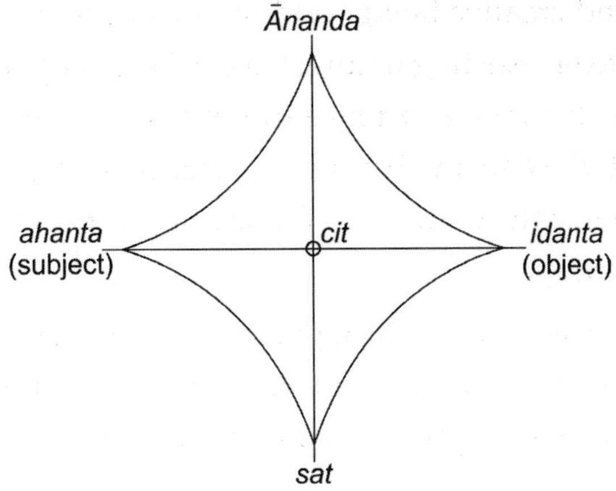

meaning or value-sense, it is *ānanda*. In every knowing, the main function of consciousness involves a value-sense also. When this *sat–cit–ānanda* becomes functional, it assumes the form of *ahantā* (subject) on the one side with its counterpart and *idantā* (object) on the other.

Due to the *māyā* inherent in the *cid-vastu* or *aṟivu*, the subject (*ahantā*) expands as a world of its own. Similarly, *idantā* also expands into a world of its own. Both these conjoined become the "world" (*prapañca jagat*).

As an individual and a living and creative being, I form an integral part of the world or nature. I am an equally living and creative reality. I and the world form one reality. So what is real in me should be the reality or substance in the being of the world also. The real substance in my being is Pure Consciousness or (*aṟivu*). Therefore, the substance underlying the being of the world is also the same Pure Consciousness. This Consciousness, owing to the *māyā* principle inherent in itself, unfolds as the world appearance. Thus, what we see as the world is *aṟivu* only, just as what we see as an ornament is gold only.

I, a living and creative being, finding my oneness with the equally living and creatively existing nature, have to be an ongoing experiential awareness. The dynamic substance in my being (called *ātmā*) finding its oneness with the equally dynamic substance (called *Brahman*) in the being of dynamic nature has to be an ongoing experiential awareness, never a static concept.

This awareness is an experience beyond words. The structural diagrams shown so far are helpful in carrying us across the realm of words to the above-mentioned experiential awareness.

There is freedom to innovate details of the structure according to one's imagination. The diagrams as such have the same importance as latitudes and longitudes on the world map. Once the intended experiential awareness is gained, the diagrams are to be discarded.

What is the essential substance in my being? It is Pure Consciousness (*a̤rivu* or *cid-vastu*) itself. It exists as the substance in my being in the guise of the knowledge that controls my being with its own structural perfection. It also exists as the substance in the being of the world as the knowledge that controls the ever-changing world, with a structural coherence of its own. In his *Ātmopadeśa Śatakam* (vv. 36-42), Narayana Guru says that this consciousness has infinite powers, classifiable as *sama* (same) and *anya* (other). The *sama* kind of knowledge perceives one Reality or *a̤rivu* alone in many appearances; and the *anya* kind of knowledge perceives the many appearances as real. The former has a vertical dimension and the latter a horizontal dimension.

Every item of knowledge should have specificity, signifying "this is such and such". There can be innumerable examples of the *anya* kind of knowledge, such as "This is a pot". Saying "This" alone does not give any specificity to the intention of one who says so. Adding the words "is a pot" gives specificity to the "This". Therefore, the "This" part of the sentence, "This is a pot" is generic and "is a pot" is its specific aspect. Any specificity can be attached to "this" creating innumerable possibilities. We can easily see that the specific knowledge concerning anything that forms part of the world is an unfolding of the generic knowledge "This".

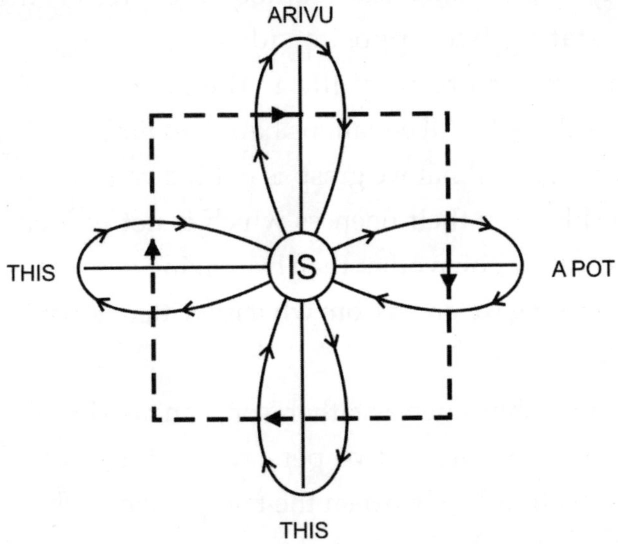

Let us now take the case of the *sama* kind of knowledge. It has only one example, "This is knowledge". Every kind of specific knowledge can easily be included in "This is knowledge". Knowledge or Consciousness is itself of a generic nature. Pondering over the generic "this" will lead us to the ultimate goal of our spiritual search (*sad-gati*).

Pointing out the structural togetherness of these two realms of knowledge, Guru says, "Everything becomes known because of the up and down spiral movements of these two kinds of knowledge". This is because without Pure Consciousness (*aṟivu*) existing, no specific knowledge of anything is possible; and without some specific knowledge, arising Pure Consciousness cannot functionally exist.

Dialectical Methodology

In our daily life, we unconsciously, and the Masters and *ṛṣis* quite deliberately, use a dialectical method to express something beyond words by stating two opposite sides simultaneously. A common example from our ordinary life is the simple algebraic equation, $(a + b)^2 = a^2 + 2ab + b^2$. All equations have two limbs with an "equalness" sign between them. What we grasp is not the value or meaning of either of the two sides, but their oneness which is not written down. It is not possible to write it down other than by making use of these two limbs. In short, the meaning we grasp from what is written down is something that is not written down.

Any relationship, such as father–son, can also be an example. When we write down or refer to two persons as "they are father and son", we grasp a relationship between the two persons, which is not actually written down or said, but which is drawn from the words "father" and "son" taken together. Here also, we intuitively grasp something not written down or said from two words said or written down taken side by side.

Take the case of two statements "crossing over death" and "attaining immortality", which really mean the same. But the statements are of opposite nature. "Crossing over" involves avoiding, which is a negative attitude for it negates something. In this case negating death. Death is a negative value, for no one likes it. Thus, the usage "crossing over death" is totally negative. On the other hand, "attaining immortality" is totally positive. "Attainment" is a positive attitude and "immortality" is of a high positive value, for everyone likes it. When these two statements of opposite nature are put together, they convey to us one and the same meaning, which is neither of the two statements. It is something neutral grasped by means of the two. This is a clear instance of the above-said method used in *Īśa Upaniṣad* 11:

> *avidyayā mṛtyuṁ tīrtvā-vidyayām amṛtam aśnute* ॥
>
> crosses death through ignorance (nescience) and enjoys immortality through knowledge (science)

This is similar to structuralism because every structure involves two opposite poles between which there is something neutral. The very same pair we noted in *Īśa Upaniṣad* 11 appears again as related to the pairs of *vināśa* (destruction) and *sambhūti* (becoming) as well. *Mantra* 14 says:

> *vināśena mṛtyum tīrtvā saṁbhūtyā 'mṛtam aśnute* ॥
>
> By knowing the destruction principle, one crosses over death, and by knowing the becoming principle, one attains immortality.

The worst destruction is death and best of all becoming is immortality. These two pairs could structurally be put together as follows:

This immortality, experienced while living, has to be an ongoing and dynamic experience of creative beingness, in which the particular beingness (*ātmā*) and universal beingness (*Brahman*) find their oneness in non-duality. This oneness is neither *ātmā* nor *Brahman*, but is a neutral

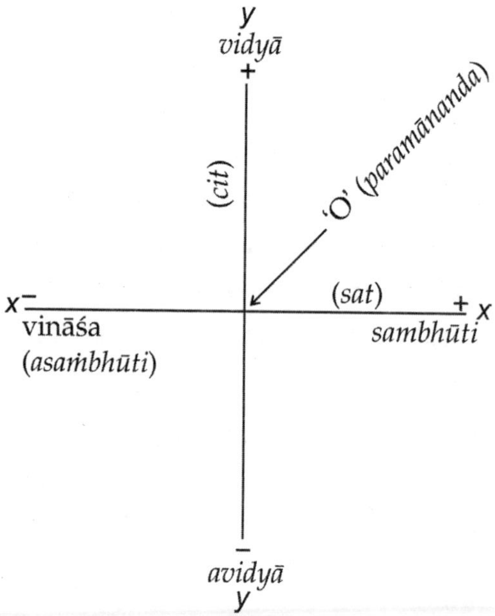

beingness which can be given expression to only as the oneness of *Brahman* and *ātman.* Attaining this neutrality beyond words is the ultimate goal of all Vedāntins.

This method of giving expression to something beyond words by pointing out two expressible opposite poles, and equating and cancelling them so that the neutral Reality beyond words could be intuitively perceived by the seeker, had been known in the West from the time of Socrates and in the East from the time of the Upaniṣadic *ṛṣi*s. In the West, it was known as *dialectics* and in India as *yoga-buddhi* as seen in the *Bhagavadgītā*, a renowned classic text of Yogaśāstra (Science of Dialectics).

4

Presidential Address

Muni Narayana Prasad

DEAR Shri Ram Nath Kovind, the honourable former President of India, all the distinguished guests on the dais, and all the dear friends assembled here on this auspicious occasion, should I feel honoured on being seated beside the honourable former President of India? No. I am a *saṁnyāsin*. I should neither feel honoured nor dishonoured on any occasion. Such is the manner in which I have been shaped by the wisdom imparted to me by my *gurus*.

What is that wisdom? It is the wisdom of myself being a ripple in the ocean of *aṟivu* (Pure Consciousness). Whatever exists in me is a fragment of that boundlessly existing Consciousness-Reality, also called *cid-vastu*.

How do we arrive at the conclusion that *aṟivu,* also called *cid-vastu,* is the one and only all-underlying Reality? Narayana Guru adopts two methods with regard to this.

The first one goes like this:

Take anything that forms a part of the world. Go on dividing it.

Finally, you arrive at a stage where nothing is visible. Still you are sure that something exists even though invisible, intangible and inaudible. That existence is absolutely known to you. The existence of this Reality underlying that one object as well as everything in general is called simple *aṟivu* (Consciousness). This is the analytical method followed by modern science also.

Another method resorted to by Narayana Guru is more logical: Take the case of knowing anything. What you really happen to know about it are its various qualities like colour, form, weight and the like. The qualified object itself *in toto* is never perceived. An image of the object, as formulated by you, gets conceived in your Consciousness. So is our knowledge about the world also. On directly perceiving innumerable qualities of this world, we conceive of the world in our Consciousness. It means that the real existence of the world is in your Consciousness, as a functional mode of your Consciousness.

Then, where do the qualities you perceive exist?

The qualities have to exist where the qualified exists. That means, in Consciousness. Understood thus, the multiplicity of the world and the oneness of Reality find their oneness in Consciousness, *aṟivu* or *cid-vastu*.

> Know that other than Consciousness no Reality does exist; this awareness ensures unitiveness.

If Consciousness alone is what really exists, what about the world we see outside? That is seen because of the creative potential called *māyā* inherent in Consciousness. *Māyā* creates the impacts that the apparent world is Real, whereas what is Real is Consciousness-Reality alone. This Reality is also known as *ātmā* and *Brahman*.

Then what about the world? Is it real or unreal? With an unenlightened person what exists is the world; and with the enlightened person what appears as the world is *ātmā* or *aṟivu* alone.

The actual existence of *ātmā* is in the form of the world. Without manifesting itself as the world *ātmā* does not exist either. What one sees around one as an ignorant individual is the world and nothing else. But, when his inner eye gets opened, what he sees around him is the *ātmā* alone; not the materialistic world! These two perceptions are entirely different from each other, while the object remains the same!

Thus, there is a non-duality between the apparent world and the Really existing *ātmā*. This is the nature of non-duality as perceived by Narayana Guru. Put otherwise, our actual life is a particular form in which the one *ātmā* functionally exists.

Guru admitted that, in principle, he followed Śaṅkara. Yet slight revaluations like the one mentioned above concerning *advaita* (non-duality) should not be left unnoticed. In fact, Narayana Guru's *advaita* is a revalued form of Śaṅkara's *advaita.* I do not go into its details here for lack of time.

Narayana Guru, as all of you know, while being an Advaitin, became involved in social problems, which is the reason for being wrongly treated by many as a social reformer. Social reforms really were an applied aspect of his philosophical vision.

What is society? No society exists outside, other than as a state of mind in individuals when looked at people collectively. As we have already noticed, nothing other than *aṟivu* has real existence. So, society also has no existence outside *aṟivu*.

Guru noticed that something wrong was happening in society; in the minds of people collectively. He then pointed it out to the people and gave them directions to set them right. His words of lofty nature, having wisdom for their basis, could not be questioned by anyone. His firm stand was not fixed in the middle of society but was in *Brahman*. He makes it very clear through the following words:

> A knower of *Brahman* is the one who, having burned down everything in the fire of wisdom, performs actions in accordance with the prescribed rules of conduct, aiming at the good of the world, while finding himself existing in *Brahman*.

There were courageous young people then to take up this message and do the needful. This caused a tremendous change in the social atmosphere of Kerala. But this did not make him a social reformer.

Nataraja Guru was his wisdom-successor. He was a man of a scientific mindset. He took upon himself the mission of giving expression to his *guru*'s vision in terms of mathematical thinking using Cartesian coordinates.

We are all familiar with a graph. It has two axes. One vertical and the other horizontal. The point where two axes meet and intersect is neither vertical nor horizontal, or it is both. Nataraja Guru considered the horizontal axis as representing the world of actualities and the vertical axis as representing the one Consciousness-Reality. Then the meeting point of the two axes represents non-duality of the two (*advaita*). With this mathematical means, it becomes easier to meditate on the mystical oneness of both *ātmā* and the actual world.

Nataraja Guru, the founder of Narayana Gurukula, wrote his magnum opus *An Integrated Science of the Absolute* as a science in which both physics and metaphysics meet and find their oneness.

Physics is one area of study in which our Consciousness functions in a particular way. Metaphysics is yet another area of study that takes place in the very same Consciousness. That means, both physics and metaphysics find their oneness in Consciousness. A science of Consciousness, should include within it both physics and metaphysics. This science is known in India as *brahma-vidyā*. This is because *Brahman* is *cit* (Consciousness) in essential content. For this reason, *brahma-vidyā* is acclaimed by the *Muṇḍaka Upaniṣad* as *sarvavidyāpratiṣṭhā*, the firm basis for all sciences.

Nataraja Guru wrote *An Integrated Science of the Absolute* in an attempt to answer this question asked to him by Narayana Guru:

Have you heard of *Kuṭṭi-cāttan*? (*Kuṭṭi-cāttan* is a mysterious invisible being that creates nuisance. His equivalent in English is Poltergeist.) He throws stones at houses. When examined, the stones are found to be collected from nowhere nearby. Does your science have any explanation for it?

No specific answer to the question can be found in the book. But the answer I derive is this: Anything is possible at the mysterious meeting point of the vertical and horizontal axes, of appearance and Reality.

The idea of "One World" with no political boundaries was very dear to Nataraja Guru. In order to promote it, he organized a "World Conference for Peace Through Unitive Understanding". It began on the Armistice Day in 1970 and lasted for eleven days.

Guru Nitya Chaitanya Yati continued as *guru* after the time of Nataraja Guru. He followed the line of thought of his *guru* and enriched it with his own psychological and aesthetical insights, and more recent developments in modern science.

As a continuator of Guru Nitya, I adhere to the methodology developed by Nataraja Guru while writing commentaries on the ten major Upaniṣads, the *Bhagavadgītā*, the complete works of Narayana Guru including his *Vedāntasūtras*, and also while writing original works.

So far, I was narrating the nature of the development of thought from Narayana Guru, our first *guru*, till now. It is my hope that it will be continued unbroken through my successors.

In 1926, the foundation stone of Narayana Gurukula in Ooty was laid by the Maharaja of Travancore Shri Chithira Thirunal Balarama Varma, and the third anniversary of the Gurukula was presided over by Sir C.P. Ramaswami Iyer, the then Dewan of Travancore. It was an event participated by VVIPs. After nearly a century, this is the first event in which a VVIP participates as a Chief Guest in the Gurukula.

I conclude my words by extending a hearty welcome to our honourable Chief Guest.

Thanks to all.

5

Guru Natarajan

Swami Tyageeswaran

THE light of wisdom, Guru, Natarajan, the complete glory of man, the form of *caitanya* (spirit) necessary for all men.

He was fixed in *samādhi*, by learning and meditating the great *mantras* (hymns) learnt from Narayana Guru.

He founded Narayana Gurukula and he dedicated his entire life to spreading the glorious message of Guru Narayana throughout the world.

He acquired wisdom without distinction of East and West, and envisioned unity of all the sciences for the good of mankind.

One World Education, One World Government, One World Economy, all these he wrote and gave to the people of the world.

From the Guru's intellect, which was equal to Himalayas, *amṛta* stream of wisdom was flowing unscathed.

No one can have the blessing of that kind of wisdom without being born with it, Guru got it naturally.

The form of the Shining World; Thou art Lord of the Shining World, in which you art one; but there is no one like you.

Guru's mother, Bhagavatī Ammā, was a Sanskrit linguist, and he also learned English from his father, Dr Palpu.

It is very rare that this great fortune is found in anyone, but in Guru, it happened by the grace of time.

As spring is reserved for the beauty of the season, you became the disciple of Guru Narayana for the welfare of mankind.

You are Guru who experienced the bliss of *samādhi* through *yoga* combined with the wisdom and practical knowledge of the *ṛṣis* (sages).

In order to unify the languages you have prepared a structure consisting of four *pāda*s (limbs). That is the secret (*rahasya*) of the iconic *praṇava* (AUM).

Guru saw that this secret was imprinted in Western thought as well. And was found by him to be the secret key to the unification of languages.

The field of language became successful with Guru and he was able to get the honour of the Royal Academy.

The unique philological skill he has employed to establish the unity of human brotherhood is equal to what happened to Ṛṣi Patañjali.

Guru is a *brahmavit* (knower of the Absolute), synthesis of wisdom and *yoga*, a person who has devoted his life to the well-being of the people of the world and its stability.

The literary realm of the world, which glorifies human society, understood them as good and poured them out in your incomparably beautiful language.

World literary poets, world famous works, shining in your heart-thread, the light of literary gem shines.

At Śakuntalā's departure, when he saw the tears of Kaṇva Muni (the Saint Kaṇva) and the poetic face of Kālidāsa Mahākavi, Guru's heart was filled with tears.

Guru's words reveal the depths of the first great light (*ādi mahas*), the *tapas* and the meditation it requires.

Guru, the shining form of wisdom and *yoga* combined, travelled all over the world to impart wisdom to the thirsty.

As the darkness of the night, turns to the dawn of the day, replacing ignorance, he always became a teacher of wisdom.

He has a heart that always thirsts for the entire human family, like the children of one mother, to live as friends.

He adopted the practice in life, philosophy and science, without any difference of East or West.

Guru, a majestic mountain-like countenance, his eyes flowing like a stream, full of merciful kindness.

Magnificent as the ocean was Guru's knowledge, and creative writing flowed from his brilliant intellect.

Art, literature and world history received the magnificent flow in Guru's autobiography.

The inner power of science, the poetic beauty, like the connection between body and soul, shined as one in you.

The first to find the guru in you was John Spears, a Western scholar who spread the message of Guru's writings.

Two people became Guru's disciples in America, the first being Harry Jacobson who founded a Gurukula there.

Swami Mangalananda, a famous orator, was his Eastern disciple who sang the nectar words of Guru into the flow of beautiful music.

Another American disciple of Guru, Garry Davis, is famous as an adventurous world traveller and a Citizen of the World.

Nitya Chaitanya Yati also joined as Your disciple and became the world renowned Guru with a loving human heart.

Guru Muni Narayana Prasad, a brilliant person of wisdom and *yoga*, leads the Gurukula today in a glorious state.

Your feet have touched the whole world many times, it is for the establishment of *gurukula*s all over the world.

Your great *samādhi* temple is full of inner light; it reveals *brahma-vidyā,* which is the basis of all sciences.

In the midst of eternal books of wisdom, in the hall of *samādhi sthāna*, which reveals your divine presence.

It is in the middle of eternal wisdom scriptures that shines as the eternal and divine *samādhi* seat of Guru.

The seat of Guru's *mahāsamādhi* is like the light of midday sun and at the same time the full moon of shedding the light of eternal peace.

6

My Life at Narayana Gurukula

Swami Tanmaya

THE Narayana Gurukula was first founded in Coonoor in 1923 by Nataraja Guru, with the blessings of Narayana Guru. Later, in 1926, it was established at the present location in Ooty. The vision of Narayana Guru is epitomized in the maxim "Man is of One kind, One faith and One God or goal". To give a living expression to this vision, Guru conceived the idea of a wisdom school, Brahmavidya Mandiram. The realization of this grand dream of the great Guru became an accomplished fact when in 1963, his disciple and founder of the Narayana Gurukula movement, Nataraja Guru, founded the Brahmavidya Mandiram at Guru Narayana Giri, Varkala, Kerala. On the eve of the laying of the foundation of the Brahmavidya Mandiram, Nataraja Guru prophesied that it would grow into a university. This prophecy came true with the formal inauguration on 19 March 1974 of the East–West Universe of Brahmavidya (Eastern Headquarters at Guru Narayana Giri, Varkala, Kerala State, India) and on 7 September 1974 of the East–West University of the Science of The Absolute (Western Headquarters in Bainbridge Island, Washington, USA).

Symphony of values is a project of the East–West Universe of Brahmavidya envisioned by Guru Nitya Chaitanya Yati. Guru wrote:

> We are interested in the general and specific interests of human beings. We classify our general interests under the following headings: Architecture, Music of Architecture, Music, Dance, Painting, Sculpting, Poetry, Theatre, Photography, Films, Folk Arts and Contemplative Sciences.

Contemplative introspection and going into silence has the culminating effect of synthesizing all these arts. This is in short "The Symphony of Values", envisioned by Guru Nitya.

In 2000, Guru Muni Narayana Prasad gave a scheme for the study of philosophy for the students of Narayana Gurukula. The three parts of the study are: Chapter 1. "Eastern Philosophy and Chinese Philosophy", Chapter 2. "Western Philosophy", and Chapter 3. "Vedanta and Narayana Guru". Main books suggested to study were *A History of Indian Philosophy* by S.N. Dasgupta and *History of Philosophy* by B.A.G. Fuller. Three periodical seminars were conducted every four months at Fernhill, Thalassery and Varkala.

I have been living in this institution since 1986. This span can be divided into three periods in my life: at Fernhill with Guru Nitya Chaitanya Yati (1986-99), 20 May 1999 to June 2012 at Fernhill without Guru Nitya and at Varkala with Guru Muni Narayana Prasad from June 2012 onwards.

Introduction

During my college days at Sree Narayana College, I used to commute along the road in front of the Gurukula which is very near to my college.

One day I along with my friends visited the Gurukula where a seminar was going on. They were discussing the high philosophy of Sartre. We could not grasp those ideas at that time and left with the belief that the Gurukula is not an institution for college students.

Later, on 19 March 1973, the college declared a holiday because Nataraja Guru, the architect of Narayana Gurukula Foundation, attained *mahāsamādhi*. So I thought that Nataraja Guru was a great Swami, as respectful as Narayana Guru and visited Gurukula to see Guru in *mahāsamādhi*.

In 1980, after completing my medical graduation from Government Medical College Thiruvananthapuram, I went into a state of mental breakdown because of the premature death of my own brother by drowning. Afterwards I was in search of meaning of life and the enigma of death and life. One of my college mates took me to the Gurukula and gave me a book titled *Mounamandahasam*, written by Guru Nitya, based on Zen stories. I became a subscriber of *Gurukulam* monthly magazine also. Then onwards I started getting an orientation of life from the Gurukula literature mainly written by Nataraja Guru on life and wisdom of unitive understanding. Most of the terminology was foreign to me when I started reading. By 1985 I finished reading all the Gurukula literature that was published till then.

I had completed a research programme in community medicine by then. The theme of the research was "Determinants of Rural Health in India" jointly by ICMR and Ford Foundation under Dr K.K. Mathan, a well-renowned statistician retired from WHO. My jurisdiction for data collection was in Kerala even though the study had two more sample areas, one each from Uttar Pradesh and Tamil Nadu. The most healthy state according to the secondary data from the Government of India was Kerala. From Kottayam, the most healthy district in Kerala, we studied 40,000 samples of the population and five samples from the rest of the population of only 20,000 each. As the death and birth rate was very low at Kottayam, we decided to double the sample population. We were not able to reach a final conclusion in this study with nineteen proformas to collect data studying 1,500 variables determining rural health. This made us think of the religious dimension in social health. The majority of families who followed Narayana Guru identifying with political ideologies rather than religious beliefs had become an enigma for me. So I decided to do an in-depth study of the philosophy of Narayana Guru to resolve the anomaly in the health development in Kerala. The density of population in Kerala is high. The economy is based on agriculture rather than industry. Educational and health development took place here without the usual favourable factors seen in the developed nations.

Guru Nitya Chaitanya Yati visited Kottayam during 1984-85 to resolve the conflict between Hindus and Christians regarding a place of worship in Nilackal. I discussed with him the enigma behind Kerala's development and the influence of Narayana Guru in the mindset of the majority of unorganized Hindus in the sample area at Kottayam. Guru gave me the permission to stay and study in Narayana Gurukula at Fernhill with him.

1986-99: With Guru Nitya in the Fernhill Gurukulam

I started serving Guru Nitya from 1986 at the Fernhill Gurukula. I was at the back and call of Guru Nitya and also doing service in the kitchen, such as washing dishes, hewing wood and drawing water. Dictations were given during prayer meetings at 9 a.m. and 7 p.m. in English and

Malayalam by Guru on *Yogasūtra* of Patañjali. I wrote down and studied them. Later, seeing my keen interest in pursuing studies, Guru gave me the responsibility of taking dictations in Malayalam and making fair copies. The routine of the Fernhill Gurukula started by 5.00–5.30 a.m. in the morning. Guru used to come to the kitchen and guide us to prepare food. Vegetables to cut were selected according to the number of inmates.

The hot water wood stove was lit by one of the *brahmacārī*s for all to take bath. The kitchen fire was lit to cook rice. Reading a book aloud was done by one volunteer while all those assembled in the kitchen to cut vegetables. Guru often led the cooking team of two or three persons. By 8.00–8.30 a.m. cooking would be over. Guru might take a morning walk with few inmates for about 1 km. At the end of the walk, before returning, Guru chanted Gāyatrī *mantra* loudly in a meditative tone. Everybody would prepare for the morning prayer. The class dictated during morning prayer was written down by a main scribe who would then read it back loudly to Guru in order to get corrected. The morning prayer could extend up to one and a half to two hours. By 11.00–11.30 a.m. everybody assembled in the dining hall to have brunch. Usually, there was no breakfast. Black coffee and biscuits were served by 5.30 a.m. while preparing food in the morning. By 3.30 p.m. in the afternoon *uppuma* was prepared by a volunteer along with a tumbler of tea and then everybody would assemble for teamwork in the vegetable garden. Guru also joined in farming often. This could continue for one and a half to two hours. One or two inmates would prepare rice gruel *kāñjī* and pulse vegetable *subjī* between 4.00–4.30 p.m. for the supper after the evening prayer between 8.30 and 9 p.m. At 7.00 p.m. evening prayer would commence and last for one to one and a half hours. Every one had free time between lunch and evening teamwork, so that they could go to town for personal purchase or other errands. Guru often rested for half to one hour after brunch. This was the routine at the time of my stay in Fernhill in 1985-86.

My role was of an errand boy, and I started by going to twenty Gurukula centres mainly to look after the Swami-in-charge. In 1987, I was sent as a bystander to Swami Muni who underwent a surgery for an infected cyst on the back of the shoulder. On 27 September 1994 Guru Nitya's mother fell down and fractured her thigh bone. She was over 90 years old at that time. She was given domiciliary care by doctors till her passing away on 27 July 1995. In 1995, Swami Bhaskarananda Giri of Vizhumala Narayana Gurukula had fracture of the femur bone (21 July). He was operated in R.V. Clinic at Palakkad and taken to Varkala Headquarters. Guru Nitya visited Swami Bhaskarananda and made all arrangements for getting the best treatment available. Other Swamis in the Gurukula had operations for infected finger and hernia. Most of them were given expert treatment by Medical College doctors.

Guru Nitya's main activity from 1985 onwards was writing books and articles in Malayalam. Guru selected the scribes from the inmates. I got the opportunity to take down *Bharatiya Manasastrathinu Oraamukham* and other essays like Mahakavi Kumaran Asan as a *sneha śilpī*, a *bhāṣya śilpī* and *svatantra śilpī*. Articles were written by Guru Nitya for Vidyabhyasa Suraksha Samiti to reduce the adverse influence of politics in schools and colleges in Kerala. Later, the articles written on education were made into a book named *Parivartanonmukha Vidyabhyasam*. Guru Nitya wrote a series of articles every week in *Malayala Manorama*, a Malayalam daily newspaper during 1993. During the end of these writings on love dialogue, Guru and a few disciples travelled from north of Kerala to south giving a series of love dialogues. These articles were printed in the book *Sneha Saṁvāda*. A series of lectures given at Kottayam Darsana Cultural Centre was published as *Pañca Darśana Mañjūṣā*. Other books written during early 1990s were *Jāti Matam Daivam* which was serialized in *Kalakaumudi* weekly, and *Manasastravum Manasastranjanmarum*, was serialized in *Manasastram* monthly. By April 1990 onwards Guru Nitya started writing himself with appropriate single line sketches for his master project on "Symphony of Values". Guru gave meditations on the *Īśāvāsya Upaniṣad* during morning classes in July 1991. Swami Ramakrishnan was the main scribe for that and the *Bṛhadāraṇyaka Upaniṣad* meditations in Malayalam. Nancy Saumya took the English dictations on the *Bṛhadāraṇyaka Upaniṣad*. By 1996 the *Bṛhadāraṇyaka Upaniṣad* (three volumes) in Malayalam were published by D.C. Books, Kottayam. Guru started Nitya Books, a publishing section at Fernhill which published many books like the translation of Rabindranath Tagore's *Gitanjali* with illustrations.

Mulbery Books of Shri Shelvy, Kozhikode, was one of the main publishers of the book *Simone de Beauvoir Avalude Kadha Parayunnu*, *Nerudayude Ormakurippukal*, etc. Poorna Books, Kozhikode, published Guru Nitya's *Vedanta Paricayam*, a commentary on the *Vivekacūḍāmaṇi* (*Crest-Jewel of Wisdom*) of Śrī Śaṅkarācārya and a few other books.

Guru Nitya not only wrote books and published but also made arrangements to sell them through book fairs. In 1994 along with Dan Arbeid from the UK, Nitya Books participated in the World Book Fair organized in Pragati Maidan, New Delhi, for two weeks. Thus Nitya Books reached the capital of India. Other book fairs in which Nitya Books participated were Calicut Book Fair (March 1994), Kottayam Book Fair (January 1995, January 1997 and January 1998), Book Fair at Sivagiri

Pilgrimage (1997 December), Cochin Book Fair (1998 January) and Book Fair at University College, Thiruvananthapuram (December 1997). I volunteered as the team leader to sell books at all these book fairs.

In 1989, three great books written by Nataraja Guru – *Life and Teachings of Narayana Guru, The Bhagavad Gita* commentary and *Autobiography of an Absolutist* – were printed by Guru Nitya under Narayana Gurukula Foundation, Varkala. At that time All India Press, in Pondicherry had the ability to print deluxe editions. The money received as a compensation from the government for the land acquired from Gurukula at Ezhimala for establishing a Naval Academy was used thus for Guruhood.

A lecture series on different aspects of Science of The Absolute were given in November 1990 at Rajadhani Auditorium, Thiruvananthapuram by Guru Nitya. Gurukula publications were sold by inmates of Gurukula for one week period, where hundreds assembled at 5.00 p.m. to attend the lectures.

Two seminars based on medical sciences were conducted by Guru Nitya at Fernhill on 8-9 April 1989, a Conference of Physicians were held in which representatives of Modern Medicine, Āyurveda, Homoeopathy, Nature Cure, Marma, Siddha and Unani participated. I delivered a welcome speech on the history of various systems of medicine under the guidance of Guru Nitya. In 1994, another seminar was hosted at Fernhill on "The Health of Women" by Guru Nitya.

In the 1990s Guru Nitya started organizing month-long Music seminars as a part of May Festival from 1st to 31st May. Every day, morning and evening Carnatic musicians from south India performed concerts.

During 10-12 June 1998, a seminar on World Government was held at Fernhill in the presence of World Citizen, Garry Davis.

From 1989 onwards Guru conducted *vidyārambha*, initiating children to learn, during Vijayadaśamī day at Thalassery Kanakamala branch of Narayana Gurukula. More than 100 children and parents assembled in the first year itself. This event continued till 1998.

Every year annual *guru-pūjā*s were conducted in all Gurukula branches. Nataraja Guru's birthday comes on 18 February and *mahāsamādhi* on 19 March. *Guru-pūjā* at Fernhill was held in the presence of Guru Nitya. A month-long preparation was done, such as going door to door collecting funds and necessary materials for feeding hundreds of villagers on that day. *Homam,* spiritual discourse by Guru, music concerts and dance or other variety entertainments were the usual programmes on *guru-pūjā.*

Every year in December, the annual Gurukula Convention is held for a week from 23 December. All Gurukula members participate. The students of Gurukula and Swamis present papers, moderate seminars and conduct prayer meetings. Guru Nitya gave his discourse after fire sacrifice at 9 a.m. More than 200 people participated and many stayed in the Gurukula and shared food. Local people participate and also contributed towards the expenses. They also shared the meals. During the evening prayer the Universal Prayer written by Narayana Guru was chanted, along with the Lord's Prayer and Al-Fatiha. Followed by that Guru gave a discourse. After supper by 9 p.m. there would be classical music concerts or dance performance and other art programmes.

A new Gurukula branch was started at Thannerpanthal, Erode, in 1988 February on the land of Shri Thankaperumal Pillai, a devotee of Nataraja

Guru. I was put in charge of the Gurukula inaugurated by Swami Muni. Guru Nitya with disciples attended the annual *guru-pūjā* on 28 February 1990. Since the land for Gurukula was not registered to the Gurukula, the functioning of that centre ceased in 1998. I spent time in this Gurukula alone during this period.

In the *Gurukulam* magazine (1987-88) my first article on "Indian Psychology" was published. Guru Nitya guided me to write this article. In December 1991 on the last day of Gurukula Convention Guru Nitya gave me the *brahmacārī dīkṣā* by giving a yellow cloth after *virāja homa*. In 1992, I presented my first paper based on *Anukampā Daśakam* (Scripture on Mercy) during the convention at Malayattoor Gurukula. In 1992, November, I gave a lecture on "Life after Death" in a seminar organized by the Theosophical Society, Thrissur. In 1994, I moderated for the first time for the seminar on education during the Annual Convention. I continued to present papers in all the later annual conventions on subjects like Bhakti, Śāṅkara Vedānta, *Darśana Mālā, Integrated Science of the Absolute* and *Ātmopadeśa Śatakam*. On 8 May 1995, Guru Nitya and friends including me made a pilgrimage to Palani hills. In 1995, Guru Nitya gave assignments to write an article to all close disciples like Jñānamayī Indira Devi and Kṛṣṇamayī Radha Devi. I got "Anatomy of Vedānta" as the subject of my essay. In 1996, a new building was built named Nataraja Guru Research and Study Library and inaugurated by Guru Nitya on 29 August at Fernhill. In the Vidyaniketan Gurukula, Guru Nitya installed a *pañcaloha* idol of Sarasvatī at the temple on the *mahāsamādhi* place of Vamakshi Amma, mother of Guru Nitya. We all participated in that function. In November 1996, Guru Nitya started a Gurukula Herbal Medicine Preservation programme at Fernhill with Dr

P. Velayudhan, BAMS, a government Ayurvedic doctor from Manjeri, Kerala. A total of thirty-three medicinal herbal plants were planted in the green house, and Guru also wrote about it in the December issue of *Gurukulam* magazine. Guru stressed the importance of preserving medicinal plants and all the Gurukula branches were to spread this message to all neighbouring households. In 1998, the meditation room in which Nataraja Guru first lived was rebuilt with wood and named as Nataraja Guru Dhyana Mandiram at Fernhill.

On 8 November 1998, Guru went for rejuvenation therapy with oil massage at Vaidyamadom Ayurvedic Centre. In November, Guru got a stroke and right-sided hemiplegia. Guru was staying in Thalassery at that time. Guru continued to stay in the Kanakamala Gurukula for a few weeks and then at a guest house of Shri Padmanabhan, at Masanagudi. Guru returned to Fernhill and attended the annual *guru-pūjā* on 21 March 1999. On 14 May, Guru Nitya attained *mahāsāmadhi* at Fernhill. The body was placed at the *samādhi* site in the woods of the Gurukula in the presence of thousands of devotees and disciples.

1999–2012: Fernhill Days Without Guru Nitya

The *yati-pūjā* of Guru Nitya was done on the 41st day of *mahāsamādhi*. I started attending seminars, presenting papers, giving talks, etc. In August 1999, I attended a seminar on "Milestones on Kerala History" at Kerala University, Thiruvananthapuram. On 4 January 2000, a talk on "The Religion of Next Millennium" was given at Rotary Club of Kottayam. On 12 January, I attended a seminar on "Suicide at Patambi College". On 13 March, I presented a paper on "The Philosophical Impact of Sree Narayana Guru" in the national seminar on "Philosophy: Yesterday,

Today and Tomorrow" at the Department of Philosophy, University College, Thiruvananthapuram. In March 2000, I attended Guru Nitya Memorial Seminar at the Department of Philosophy, Sree Sankaracharya University of Sanskrit, Kalady. On 13-14 November 2000, I presented a paper on "Renaissance and Narayana Guru" in the All India Seminar on Guru Nanak, Sant Kabir and Sree Narayana Guru, organized by Department of Philosophy, University of Calicut. In the May issue of *Gurudevan* magazine, I wrote an article in remembrance of Guru Nitya, "*Atmavil Urangunna Snehasmitam*". On 12 November 2001, a paper was presented on "Social Mobility and Narayana Guru", in the Seminar on the Contemporary Socio-economic and Cultural Condition of Sree Narayana Guru, held by Department of Philosophy, Osmania University, Hyderabad. In 2002, I presented a paper at Dravidian University, Kuppam, Andhra Pradesh on the "Universal Vision of Sree Narayana Guru". In 2009, in a national seminar on Gandhi, Science and Environment, organized by Gandhian Institute, Wardha, Maharashtra, presented a paper on "Environment and Spirituality". In the same year, another paper was presented in the seminar on Sree Narayana Guru, at Potti Sreeramulu University, Hyderabad on the subject "Blending of Spiritual and Temporal in the Sanskrit Compositions of Narayana Guru". On 25 January 2011, I presented a paper at the Institute of Gandhian Studies, Wardha, a paper on "Svadesi and Religion" in the seminar on Gandhi, Svadesi and the Present Economic Crisis. In the same year, a paper on "Holistic Education" was presented in the seminar on Education at Sree Narayana Training College, Nedunganda, Thiruvananthapuram. During 7-13 June 2011, I presented a paper on "Methodological Innovations of Narayana Guru in Metaphysical Inquiry" in the conference on "The

Philosophy, Vision, and Works of Narayana Guru as an Instrument of Egalitarian Change" at Indian Institute of Advanced Study, Simla.

Guru Nitya remembrance meetings (*smaraṇāñjali*s) were held at Narayana Gurukula, Thrippunithura (5 September 1999), Grace Cottage, Edapally (18 September 1999) and Vidyaniketan, Konni (14 May 2000).

On 19 May 1999, I attended a Workshop on Nature Cure at Kottayam along with Dr George David, a macrobiotic practitioner and Dr Rajendra Babu, a Modern Medicine and Ayurvedic practitioner. Both these doctors were friends of Guru Nitya. In May 2000, a discussion class was arranged at Fernhill Gurukula, on "Health Care, Traditional Medicines and Medicinal Herbs". Rev. Fr. Joseph Chittoor, Swami Vinaya Chaitanya, Swami Saradananda of Gudalur branch of Narayana Gurukula and Indu Malla, a poet and writer from Ooty, participated. Another workshop was held at Fernhill on "Flora and Health" with an exhibition of medicinal plants on 25 April 2001. On 8-9 August 2001 a workshop was held in Kanakamala Gurukula, Thalassery, Kerala led by Rev. Fr. Joseph Chittoor and the local traditional Ayurvedic practitioners called *vaidya*s. In the following two days the "Flora and Health" workshop was held at Narayana Gurukula, Kottekkad, Cheruvathur, Kasaragod district, Kerala. On 6 January 2002 I underwent a six-week training on Herbal Medicine at The Catholic Hospital Association of India Centre at Secunderabad, Andhra Pradesh under Rev. Fr. Joseph Chittoor. A think-tank was held at Fernhill on "Biodiversity and Herbal Medicine" on 1 June. Two more herbal medicine classes were held at Vidyaniketan, Konni and a Church at Padathalamoodu, Tamil Nadu, with Rev. Fr. Joseph Chittoor, during 5-8 September.

On 8 December, a paper on "Energy Medicines" was presented with Dr George David at the Pranic Healing Centre, Kottayam. On 6-7 February 2003, a Herbal Medicine and Macrobiotics workshop was held at Narayana Gurukula, Kundaliyur, Thrissur guided by Rev. Fr. Joseph Chittoor and Dr George David. On 13 August 2003, another Herbal Medicine and Traditional Toxicology workshop was held at Guru Nitya Chaitanya Yati Memorial Library and reading room at Tiruvottiyur, Chennai with Rev. Fr. Joseph Chittoor and Shine Vaidyar. In 2004, on 25-26 January, a Macrobiotic Workshop was held with Dr George David at Fernhill. In February I attended the Organic Farmers Meet at Tapovan, Erumely, Kottayam organized by Shri Suresh Babu Malloor. I underwent training in Snakebite management for ten days under Dr Ramachandra Aithal at Puttur, Mangalore. On 26-27 November, a two-day Herbal Medicine workshop was held with Rev. Joseph Chittoor at Koodaranhi Church, Kozhikode. On 6 January onwards, a year-long monthly workshop for twenty traditional healers was held, led by Rev. Joseph Chittoor, Shri Shanmughan Vaidyar, Shri Shine Vaidyar and Br. Sivadasan. On 13 January, Herbal Medicine classes were held by Rev. Fr. Joseph Chittoor at Boys Town, Mananthavady and also at Janasoukya at Mananthavady. During 21-23 February, Herbal Medicine classes were held at Madikeri, Karnataka by Rev. Fr. Joseph Chittoor. During 20-23 March, Herbal Medicine workshop was held at Pykara Hans Vihar Ashram for tribal students in Ekal Foundation. On 20-21 January 2008, classes were taken on Holistic Medicine at Vanamoolika, Pulpally along with Rev. Fr. Joseph Chittoor. On 18-19, a get together of traditional healers was held at Gramaniketan, along with Swami Charles Chaitanya and Swami Vidyadhiraja and Rev. Fr. Joseph

Chittoor. Gramaniketan is a self-help group of women organized by Shri Jose, a follower of Guru Nitya and Shri Sabu of Thoppil Vaidya family. On 20-21 January, a two-day Herbal Medicine workshop was held in a library at Pandipalli, Muvattupuzha, the native place of Swami Vinaya Chaitanya. On 8 October 2008, a herbal medicine class was conducted at Shimoga by Rev. Fr. Joseph Chittoor. On 1 September, a herbal medicine class was held at Vallabhassery polyclinic by Rev Fr. Joseph Chittoor. The class was organized by Dr Sathishkumar and Dr Jayanty. On 8 November, I attended the Health Camp at Kadakkarapally, Cherthala.

From 1999 to 2012, the morning and evening classes were held at Fernhill with *homa* on Sunday morning at 9.30 a.m. We started reading and discussing the books written by Nataraja Guru, Guru Nitya and Guru Muni based on the philosophy of Narayana Guru. *Life and Teachings of Narayana Guru*, *Autobiography of an Absolutist*, *Bhagavad Gita* commentary of Nataraja Guru and other books like *Dialectics*, *One World*, *Psychology of Darśanamālā*, *An Integrated Science of The Absolute* and *Saundaryalaharī*, commentary by Guru Nataraja and Guru Nitya, commentaries of the *Īśāvāsya Upaniṣad*, *Bṛhadāraṇyaka Upaniṣad* and others such as the *Yoga Darśana*, *Vivekacūḍāmaṇi* and *Yoga-Vāsiṣṭha*. Study of *A History of Indian Philosophy* by S.N. Dasgupta and *History of Philosophy* by Fuller were discussed. We used to chant Universal Prayer in the morning and compositions of Narayana Guru, the Upaniṣads, the *Bhagavadgītā* and *Saundaryalaharī* in the evening.

During the month of May, a Tamil Poets Meet was held at Fernhill from 2000 onwards under the leadership of Jeyamohan and Nirmalya Mani. Both are followers of Guru Nitya. Every year Nancy Saumya of Bainbridge Narayana Gurukula came and took Meditation classes at

Fernhill and Kanakamala branches for two days each. So also Swami Charles Chaitanya conducted Yoga classes in Ooty. E-commerce classes by Shri Radhakrishnan, MBA were arranged at Fernhill on 1 May 2004 and at Tripunithura Gurukula on 13 June.

Mahasamadhi Mandir of Guru Nitya commenced construction on 1 June 2001 and was inaugurated on 11 May 2003 by Guru Muni Narayana Prasad. On 16 March 2003, the *Autobiography of Guru Nitya* was released in Thrissur.

Guru Muni gave me *saṁnyāsa* on 1 January 2002 and gave me the name Swami Tanmaya. I joined Guru Muni for giving educational initiation on Vijayadaśamī at Kanakamala Gurukula from 2002 onwards.

I made a visit to the villages of Orissa (now Odisha) with a *brahmacārī* from there for a week during October 2001. On the same year in November, I went to Mount Abu, Rajasthan to participate in a religious conference organized by Brahma Kumaris. From 2003 onwards I attended the annual *guru-pūjā* at Valley View Ashram of Swami Devi Dayanand, a disciple of Nataraja Guru from the Badaga community of Ooty. Guru-Pūrṇimā and Śaṅkara Jayantī were celebrated there in the traditional manner. New year meetings on the eve of 31 December were celebrated in the Meditation Centre at Malayattoor with Swami Charles Chaitanya and Dr George David in 2003 and 2004. In 2002, I attended an interreligious meeting at the Renewal Centre at Neriamangalam. I visited a Jungian psychologist, Dr George Ammattil at Kannur and discussed about Jungian psychology of unconscious.

I had started collecting information on temples related to Narayana Guru for the book written by Shri K.P. Balakrishnan of Nitya Books,

Kannur. Guru Nitya deputed both of us to write a book on temples with beautiful photographs. A few other books of Guru Nitya released after his *mahāsamādhi* were *Parisuddha Quranu Hrudayanjali, Ente Hrudayathile Gandhi* and *Rogatheyum Maranatheyum Suhruttukkalakkaam.*

A few Swamis attained *mahāsamādhi* from 2000 onwards. Swami Sree Surya Narayana Dikshitar, a disciple of Guru Narayana, Swamini Radhananda Giri of Sreekrishnasram, Villvarani, Tiruvannamalai (2001) with whom I had a bipolarity from 1993, Swami Bhaskarananda Giri (2002) and Kumaraswamy of Mangalabharathy (2004), a disciple of Nataraja Guru, Swami Chidrupananda of Sree Narayana Sevasram, at Kāñcīpuram, a branch of Sivagiri Mutt (2006).

I started participating in all annual *guru-pūjā*s in the *gurukula* branches and conventions regularly. Narayana Guru's Jayantī and *mahāsamādhi* were celebrated every year in all Gurukula centres. I also arranged classes of senior teachers of Gurukula, like Swami Chidambara Theertha and Professor Ramachandran. Monthly classes were started from 4 February 2004 at Oachira Gurukula by Professor Ramachandran.

In 2004, I went for a pilgrimage to Haridwar, Rishikesh, Gangotri and Uttarakasi for a month. On the last day of Gurukula convention other participants and I started visiting my mother staying about 5 km away from Gurukula, from 2004 to 2015.

From 2004 onwards, I visited Hyderabad and stayed in the open house for feeding students from low-income group, organized as a social intervention programme of life health reinforcement managed by Dr Suryaprakash Vinjamury. From 2004 onwards, I went to Entayar to give spiritual support to a classmate Dr Sukumaran who had suffered from multiple myeloma.

In 2005, some prayer classes were arranged in the neighbouring villages with local devotees of Guru Narayana. I have participated in the monthly Chataya Puja celebrations in Sree Narayana organizations in Ooty and Coimbatore. I also participated in the annual *guru-pūjā* at Hans Vihar Ashram at Pykara, Ooty. In 2008 and 2009, I attended the children's camp at Mangalabharathy Ashram for women at Kalady. I also started participating in the *Yatismṛti* after the *mahāsamādhi* of Guru Nitya organized by Professor Madhavan Nair of Adoor. I was invited to the Illathu Pillaimar Sangham meet at Tiruppur on 23 August 2008. From September 2008 onwards I started visiting Nagpur. I had participated in the annual Chataya Puja of SNDP branch, Nagpur. Later I went to the Gandhian Institute at Wardha to present papers in the seminars based on Gandhian philosophy conducted by Dr Siby Joseph, an admirer of Guru Nitya. In 2012 January, I participated in the *guru-pūjā* at the Santalinga Swami Ashram at Thiruparankundram, Madurai, Tamil Nadu.

From 2008 onwards, regular weekly classes were conducted at Trippunithura and Edappally branches of Gurukula based on *Ātmopadeśa Śatakam* and other compositions of Narayana Guru till 2012.

A new three-storey building was inaugurated at Varkala Narayana Gurukula headquarters on 22 November 2008. Now I am staying in that building. On 14 May 2009, the 70th birthday of Guru Muni was celebrated by Gurusmarana Samiti, Ernakulam. Shri T.S. Siddharthan, a disciple of Guru Nitya arranged this function.

On 3 May 2010, the repair of the kitchen building at Fernhill started and finished by June 2012. I have shifted my stay to Varkala headquarters from 2012 June onwards.

2012 to 2019 at Narayana Gurukula, Varkala

Now the bipolarity with the Guruhood became intimate with Guru Muni Narayana Prasad because of the stay at Varkala. Here the routine was different from Fernhill. The morning prayer is at 9 a.m. by chanting *Daivadaśakam*. During the evening prayer along with *Daivadaśakam* we chanted Lord's Prayer and Al-Fatiha. Every Sunday, morning at 9.30 we had fire sacrifice and discourse by Guru. My duty is now relieved from managing a *gurukula* branch. We have an office here at Varkala for administration and book sales. A printing press is also attached. So one of the main jobs is editing and proofreading. Copying the books of Guru Nitya was my major duty with Guru Nitya. Here we have DTP and printers so that office work became less tiresome.

I started travelling in and out of India by November 2012. On 11 November in Tirupaty Padmavathy University the Telugu version of *Nārāyaṇasmṛti* was released. I made the presidential address there. On 25 November at Bangalore Malayalee Samajam I gave a talk on "One World Vision". In January 2013 I was awarded Rs. 50,000 from T. Raghavan Trust at Thrissur for my services to spread the message of Narayana Guru. From 14 February onwards, I started an all-India pilgrimage with the sum awarded. First I went to Agra, then to Chitrakut. Then went to Allahabad to participate in the Kumbha Mela. In April I reached Varanasi, Kāśī and stayed at Malayala Swami Ashram with Swami Omkarananda. I visited Kāśī Viśvanātha Temple and Banaras Hindu University Campus. On 21 April, I participated in the birthday celebrations of Malayala Swami with 1,500 followers of Swami from Andhra Pradesh. I visited Rev. Sister Pauline staying at Sāranāth and the Buddha centre at Sāranāth. On 25 onwards I stayed at Hanumān Temple at Kāśī chanting *sītārām* for three days. Then I went to Mughalsarai Railway Station and left for Ahmedabad and stayed at Sree Narayana Cultural Mission guest house. I reached Dwaraka by 1 May and stayed at Bharat Seva Ashram Sangh. I visited Somanātha Temple and then left for Baroda. I reached Hans Vihar Ashram on the bank of River Narmada. On 8 May, I reached Mumbai and stayed for a week at Sree Narayana Guru Mandira Samiti. I reached Kolkata on 20 May and stayed at Sree Narayana Seva Sangham. I reached Shillong on 26 and stayed with Shri Mohan of Sree Narayana Cultural Centre. By 1 June, I reached Guwahati, Assam and stayed with Mr Harish Ram at Santigiri Ayurveda and Siddha Hospital.

On 6 June, I reached Ujjain and visited the holy places there. On 10 June, I reached Bhopal and stayed at the Sree Narayana Mission there, also attended the Chataya Puja meeting at Sree Narayana Sandesa Centre, Kolar. On 19 June, I visited Jaipur Sree Narayana Dharma Seva Samiti and participated in the monthly Chataya Puja. I stayed with Mr Madhusudanan in his hotel. Then I went to Raipur and met Mr Ajith Ramachandran of Clipper on 28 June. I visited the Śiva Temple, managed by Sree Narayana Sandesa Samiti, Raipur, Chhattisgarh. I took a class at the Sree Narayana Guru Vidya Bhavan, Bhilai for plus two students. I reached Jammu on 30 June and visited Vaishno Devi Temple there and later reached Mathurā, the birthplace of Śrī Kṛṣṇa. I reached Patna by 8 August and visited Gayā and Nālandā. On 12 August I reached Pune and stayed with Shri Regunandan of Sree Narayana Guru Samiti and later at Guru Mandira Temple at Dehu Road. Thus ended my all India pilgrimage.

In August 2014, I left for the UK as an invitee of Sree Narayana Guru Mission, UK. I stayed in the residence of Shri Kochugovindan. I participated in Guru Jayantī celebrations in three centres of the mission at Eastham, Croydon and Southall. On 8 December 2014, I returned to Varkala. With a few serious students of Narayana Guru's philosophy from the UK, we started a Skype class, studying the works of Nataraja Guru, such as *Word of the Guru, Autobiography of an Absolutist, One World* and *Dialectics* for five years. Sanal Madhavan, Subhash Sadasivan, Saleena Sadasivan, Shyla Anandan, Angelina, Praveen and Sudha Sasidharan were in the group. On 19 October 2017, I spent one week in Kuwait as a guest of Saradhy, a group of devotees lead by Shri Ajayakumar of Konni. I participated in the Sree Narayana Guru *samādhi-pūjā* and various classes in different units of Saradhy.

Classes on herbal medicine and lifestyle diseases continued with Rev. Fr. Joseph Chittoor in different parts of Kerala, Tamil Nadu, Karnataka and Andhra Pradesh. On 13 January 2013, a herbal medicine class was held in the Church at Kelakam, Kannur district. In the month of September, I went to Jeeva Dharu at H.D. Kotta Mysore, Karnataka. Rev. Fr. Joseph Chittoor advised me to go to Paduvakote Gurumandiram run by Sree Narayana Guru devotees and give a class on Guru's vision and health on 13 September 2013. Fr. Joseph Chittoor and I went to Shimoga in October and conducted health classes in the Catholic Church there. In 2015, we took a series of classes for an NGO run by Shri Varghese at Arsikere, Karnataka. On 6 October 2016 herbal medicine classes were conducted in Peravoor and Nedumpoil for farmers. On 2 January 2017, cancer awareness classes were taken at Mananthavady. How to avoid a relapse of cancer in a person through proper diet and herbs was the theme. From 22 January onwards, I participated in the seed festival organized by organic farmers group under the leadership of Rev. Father

Joy. Rev. Fr. Joseph Chittoor and I talked on the importance of organic farming. Next year also, I had participated in the seed festival at Payyanur, Kannur district and gave a talk on "How to make farmers scientifically literate". On 16 March 2017, we conducted herbal medicine classes at Thrissilery, 16th mile Puthussery Kadavu and rehabilitation centre for mentally challenged run by Shri Johnny at Vazhavatta in Wayanad district, and on 21 and 22 at Narasimharajapura near Sringeri Mutt. I visited the Sarasvathī Temple there in the evening. On 17 May 2017, took a herbal medicine class organized by Guruvarul Study Group, Varkala arranged by Mr Muthana Thaha, a follower of Nataraja Guru and Guru Nitya. On 18 July, a class on lifestyle diseases was taken along with Rev. Fr. Joseph Chittoor for the women's wing of Wayanad Social Service Society, an organization of organic farmers. On 21 July, a class was taken at the church at Iritty, Kannur. On 21 November morning, a class on lifestyle diseases was held in the church at Kelakom and afternoon at the SNDP Kudumba unit at Kottiyoor. During the month of November, two classes were taken on lifestyle diseases at Dharmagiri Church, Manvayal and SNDP at Nandhatty, Gudalur, Tamil Nadu. In 2018, a herbal medicine class was taken at Maranalloor, Neyyattinkara under the NGO run by Dr Guru Vamsy and Dr C. Bhaskaran.

I continued to visit open house of Hyderabad once in a year. I presented papers in Varkala Gurukula annual convention. I participated in the *samādhi-pūjā* of Swamini Radhananda Giri on 29 January every year from 2001 onwards. I attended the celebration of fifty years of Pastoral life of Rev. Fr. Joseph Chittoor at Boys Town, Mananthavady on 11 January 2014. On 11 September 2015, I participated in the idol installation ceremony at Pacha, Chekkidikkad, Alappuzha district. On

18 April, Dr Sathishkumar Vallabhassery and I visited forty households in Kuttanad, organized by Guru Dharma Pracharana Sabha, Sivagiri Mutt. In June 2015, I attended a three-day youth camp coordinated by Ms Manju Sasidharan in Goa. The programme was hosted by Sree Narayana Guru Memorial Society, Panda, Goa. I attended the inaugural ceremony of Clipper 28, an Internet Newspaper in Hindi and English from Raipur, Chhattisgarh on 19 August 2015. On 28 August, the Telugu version of Guru Nitya's commentary of *Daivadaśakam*, a universal prayer for humanity was released. The programme was hosted by Sree Narayana Darsana Vedi at Visakhapatnam, Andhra Pradesh. The translation was done by Shri L.R. Swami. The programme was initiated by Shri Mohankumar. On 13 February 2017, I participated in the jubilee of the pastoral life of Rev. Sister Rose, a herbal medicine trainer at Anchalummood Church, Trivandrum district. On 30 May 2017, I participated in the marriage ceremony of low-income group Hindu couples to bless them at Yathim Khana, Moottil, Wayanad district. A two-day class was taken for the Sree Narayana Guru Study Centre at Chempazhanty for the trainees on 16-17 June 2017. On 1 February 2018, an inaugural speech was delivered in the zonal meeting of delegates of different units of Guru Dharma Pracharana Sabha, Kottayam. On 22 April, another speech was given in the meeting of Kudumba unit members of SNDP at Puthuppally, Pampady, Kottayam. On 14 May morning, I was invited by the Dooradarsan Kendra, Thiruvananthapuram to a live programme on Guru Nitya.

On 10 January 2019, I was invited to observe the monthly Chataya Puja at the seven *gurumandirams* in the SNDP branch at Vandanam, Alappuzha. On 12 January, I was invited by a Kudumba unit at Murinjakal to talk on Guru Nitya for their anniversary. I participated in the birthday celebration of Swami Bodhananda at his birthplace, Chirackal, Thrissur. On 16 February, I attended the function of the release of commentary of Holy Quran by Muhammed Mustafa Moulavi. On 14 December, I spoke in the book release function of the novel, *Niyogam* by Shri Syamlal at Chirayinkeezhu, Sabhavila.

Monthly classes at Angamoozhy SNDP and visiting all the fourteen family units started on 10 May 2015 and continued till 5 May 2018. Youths were taught to chant *Sampūrṇakr̥tigal* of Narayana Guru and Gurukula *Dhyānamañjūṣā*, a collection of relevant parts from the Upaniṣads. An online class, started in the name of Guruvarul Padana Sangham, Angamoozhy, completed Nataraja Guru's *Word of the Guru* and commenced *An Autobiography of an Absolutist* every fortnight. It is a one and a half hour long class. Out of the ten participants one person presides over the session and one member presents the summary of five to six paragraphs from the book. Other members also contribute to clear all doubts through discussion. Sree Narayana Guru Study Centre in Karukachal, Kottayam also started functioning from 10 May 2015 under the management of Sr. Ravindra Nath. Monthly classes were conducted by various teachers on English Grammar, the *Bhagavadgītā*, *Ātmopadeśa Śatakam* and *Darśanamālā*. I was also associated with the study programme till 4 April 2018. Shri Hariharan Unnithan translated compositions like *Ātmopadeśa Śatakam* of Narayana Guru into Hindi. We had a Translators' Meet on 9 October 2017. In 2018, a new study group was formed at Mutharikalam, Payyavoor, Kannur district to study Guru's vision. It continued till 2019. Books on Guru were given and elders taught children every Sunday morning. On 7 May 2018, another study group was organized in Kumily, Amayannoor (6 May 2019) and Chathenkary, Thiruvalla (18 September 2018) and

Nellippara, Idukki (10 June 2019). I stayed in Gurukula at Oachira and Vizhumala alone for contemplation during January and February 2016.

Many members of the Gurukula movement died in this period from 2012 to 2019. On 30 March 2014 Swami Shanmugha Narayanan, a close associate of Nataraja Guru, at Thrissur; Swamini Jñānamayī Indira Devi at Palakkad; on 15 December, Muthana Thaha at Varkala, on 3 March 2019, Professor Ramachandran and Shri Suthan at Thiruvananthapuram, etc.

I attended a seminar on Narayana Guru Philosophy, held at Brahmavidya Mandir, Varkala organized by the Indian Council of Philosophical Research (ICPR), New Delhi on 5-6 May 2016. On 15-16 October, I presented a paper on "One Universal Religion for All Human Beings" at the Seminar on Interfaith Harmony at Sri Venkateswara University, Tirupati, Andhra Pradesh. On 2 February 2019, I was invited to talk on Guru's Vision at the annual meeting of Sree Narayana Darsan Prerana Sangh (Karnataka). On 23 May 2019, I was invited to give a talk at Jyotisastra Samgham annual meet at Haripad, Alappuzha. From 19 June onward, for two weeks, I stayed in meditation at Divya Santiniketan, Malayattoor, Ernakulam.

These are my experiences at Narayana Gurukula Foundation from 1986 to 2019. My main study was on the *Integrated Science of The Absolute* (*brahma-vidyā*). For Guru Nitya, I did services (*guru-sevā*). During the period after Guru's *mahāsamādhi*, I did service to Guru Muni first staying at Fernhill and later staying at Varkala. The main *sādhana* I had with Guru Muni was associating with the translation of *Integrated Science of The Absolute* and sharing the wisdom from it by writing articles in *Gurukulam* monthly. More in-depth understanding of Narayana Guru vision is happening slowly.

The classes on health taken by me include herbal medicine, home remedies, healthy dietary practices from the traditional medicines such as Āyurveda and macrobiotics, holistic health, palliative care for cancer patients with food and herbs like aloe vera. Other topics include lifestyle diseases prevention through proper diet, regular exercise, environmental and ecological sensitivity, stress reduction and self-responsibility which are the main pillars of Āyurveda. Organic farming and importance of scientific literacy of farmers are other areas of interest. For the classes on Narayana Guru, I consider *vaidya* for the problems relating to body, *vaidika* for the mind as well as Yogaśāstra and *brahma-vidyā* for the realm of Self (*ātmā*). Literature available on herbs and home remedies prepared by Rev. Fr. Joseph Chittoor and Rev. Sister Innocent are mainly used to teach home remedies. For macrobiotic classes the food patterns and home remedies are taught with the help of a book on macrobiotics by Dr George David. For the classes on the vision of Narayana Guru, Gurukula publications like *Sampūrṇakṛtigal* of Narayana Guru and *Gurukula Dhyānamañjūṣā,* a collection of relevant sections from the Upaniṣads, are chanted to the participants with explanations.

The way of life in the Gurukula is found worth living for me amidst all the challenges of horizontality in *gurukula* life. When too much horizontal winds affected my life, I left without telling anyone, even Guru Nitya, and went for a pilgrimage to a hill near Tiruvannamalai and returned after fifteen days as I planned. The wandering without money and with limited food received as *bhikṣā* occasionally with the Grace of God changed my view on life and death. I was doing *sādhanā* with *śrīcakra* and when I finished one round of chanting, I was taken to an illusion and frenzy. It happened on 6 December 1993. Another blow of horizontal forces hit me on 12 June 2012 at Fernhill. It was another encounter with life and death experience from some unknown people related to Gurukula at Fernhill. Afterwards I shifted my stay to Varkala to take rest for four months to normalize myself. Both brother Girish and Yogi Pramod helped me to overcome my debilities. They were serving the Fernhill Gurukula and they studied *brahma-vidyā* and Yogaśāstra one year before the untoward incident from horizontality happened.

I continued to stay in this wonder world of Narayana Gurukula till the time left to live on this blue dot, called earth.

7

The "Frame of Reference" of Narayana Gurukula

P.K. Sabu

> To write a book about Guru, to found a centre in the West for the teaching of *brahma-vidyā*, to lecture in the United States on One Religion, to complete still further my education, both Eastern and Western, and to study the programmes of the United Nations and the UNESCO so as to be able to form ideas about world unity, were the main items of the five-point birthday resolution that I made for myself on my fiftieth birthday.

THIS is a passage from the autobiography of Nataraja Guru. Being a self appraisal of the founder of Narayana Gurukula, these words of Nataraja Guru serve as a beacon for anyone who wants to understand the "frame of reference" of Narayana Gurukula. A frame of reference is basically a bound space. It can be conceptual or physical. Like the "frame" of a photograph or hanging painting, the frame of reference can define and limit a field. Also, like the "frames" of a building or the rigid skeleton as a "frame" work of the human body, it is supportive too. So what exactly is the frame of reference of Narayana Gurukula, its scope and limits at the organizational and philosophical domains?

A few more quotes from Nataraja Guru will make it easy to fix the issue precisely. On 1 January 1956 he made a declaration. This declaration forms the first part of the Prospectus of Narayana Gurukula Foundation, which can be considered as the "Constitution" of the Gurukula. This is the opening paragraph:

> This is to let it be known generally, and in particular to those interested in the Gurukula Foundation, that I, Dr Natarajan (also known among my disciples as Nataraja Guru and more plainly to be known as Natarajan) being the Head and Founder of the Narayana Gurukula Foundation, Varkala P.O., T.C. State have this 1st January 1956 overtly adopted *saṁnyāsa* (pattern of life of a man of renunciation) as understood in the Indian contemplative tradition. This is in spite of the fact that I have since 1923 been a *tyāgī* (relinquisher of benefits of effort) and even received in 1926 the yellow cloth from Guru Narayana expressly given to me in recognition of my life-long dedication to Guru-cause. The *saṁnyāsa* (renunciation) at present assumed, after crossing my sixtieth year of life, is meant to confirm the *tyāga* (relinquishment) and to be a more open and objective re-dedication of the same line of detachment in a mature form, as implied in the *Bhagavadgītā* (XVIII.49) and as laid down and permitted in the *Narayana Smriti* (Code of Life written under the direction of Guru Narayana, verse 131 *Sri Narayana Dharma*, p. 56).

In formal terms, this declaration can be considered as a preamble of the Gurukula Memoranda. The preamble of all formal documents sets the philosophical basis to interpret the remaining parts of it fully in accordance with the purpose for which it had been written by the founders of the institution, for which the document serves as its constitution or guideline. The purpose for which it is written is to make sure that the future course of the institution will be along the right path, as visualized by the founders. The declaration of Nataraja Guru too can be seen in that spirit. Since it contains certain autobiographical references, it can also be seen as the "frame of reference" of his own life.

From the frame of reference of the life of its founder, chapter two of the prospectus, the Gurukula Memorandum, leads us to the frame of reference of the organization founded by him. This chapter throws light on how to fix the exact frame of reference of the Gurukula without confusing it with the prejudices inherited from the acquaintance of frame of references of other organizations which carry the title "Sree Narayana" in their names. The "Sree Narayana" organizations, by and large, look at Narayana Guru only through the frame of the Kerala history of yesteryears. In a way it is the frame through which society "sees" Narayana Guru. Contrary to this narrow perspective, Nataraja Guru "observed" Narayana Guru through a global frame. The explanation on the name and origin of Gurukula, given in the prospectus, itself is self-explanatory to give clarity on the difference between how society is "seeing" Narayana Guru and how Nataraja Guru "observed" his own Guru. We read:

> The "Narayana Gurukula", also hereinafter referred to more shortly as the "Gurukula" or "Gurukula Foundation" or the "Foundation" is the name of an institution conforming to what is well known in the traditional life of India from most ancient times and which is also sometimes referred to as "Mutt" or "Ashram". It consists of a *guru* (spiritual preceptor) and *kula* (family consisting of his disciples who together constitute a spiritual family) headed by the teacher of wisdom. In the present case, they together follow the teachings and conform to

> the way of life represented by Guru Narayana of South India (1854 to 1928) who in his own life time taught contemplative wisdom recognized as Advaita Vedānta (non-dual philosophy) or *brahma-vidyā* (Science of The Absolute) which is not other than *ātma-vidyā* (Self-knowledge) conforming to the spirit of the Upaniṣads and following the general lines of the teachings of Śaṅkarācārya (788 CE) though in a somewhat revised and revalued form. Thus the foundation, unlike the "Trusts" known in English Law, conforms to a special type of institution particularly belonging to the Indian background from most ancient times.

A scrutiny of this passage written by the founder reveals that both the organizational and the philosophical frames of Gurukula are implied in it. An important organizational frame is that it is not founded according to English Trust laws, it follows the traditional Indian custom of *paramparā* (vertical hierarchical line of succession). This is further clarified in the front cover of the prospectus by way of defining Gurukula as

> a Guru–disciple foundation according to Indian (Hindu) tradition based on *paramparā* or vertical hierarchical succession for teaching unitive understanding and universal brotherhood ....

The choice of *paramparā* over the English law has a specific reason too. It can be seen as a reflection of the commitment of Nataraja Guru to see that the intentions of his own Guru should be properly respected and carried over to generations. This succession method was clearly envisaged in the Will and Testament written by Guru Narayana. Unfortunately as a result of a scheme suit filed in a court of law by some members of an organization existing in the name of Guru, the tradition of one-person succession became invalidated. Nataraja Guru might have been pained and shocked by this move against the will as well as Will of Guru. There is reason to believe that it is the pain and disappointment induced by this move that urged him to make it sure that there is at least one organization to respect Guru Narayana's will in letter and spirit and hence the *paramparā* system incorporated in Gurukula. This inference is not a wild guess. It has the support of Nataraja Guru's own words:

> Many existing organizations at present in the name of Guru give only scant respect to this feature. Hence the justification for a proper Gurukula constituted in keeping with the Will and intentions of the original Guru Narayana.

It is obvious that an "existing organization in the name of Guru", referred by Nataraja Guru, echoes certain developments in Sree Narayana Dharma Sanghom Trust during the 1950s, which culminated in the invalidation of Guru Narayana's Will.

At the organizational level, the Gurukula has one more Indian frame to support its existence. It is the Constitution of India. This legal affiliation had been made explicit in Section 4 of the Gurukula Memoranda. It states:

> The Narayana Gurukula is a foundation that derives its status in law from Article 19 of the Indian Constitution by which in 19(1) it guarantees freedom to acquire, hold and dispose of properties involving concrete and abstract rights.

Further, Articles 25 and 26 are also quoted as the legal premise that guarantees the freedom to run a foundation like Narayana Gurukula which belongs to the category of a Mutt-like institution. The foundation is also conceived as a "Charitable" and "Public" one which, in India, is referred as a *dharmasthāpanam* (foundation for the advancement of a righteous life).

Section 5 of the Gurukula Memorandum is "The Gurukula Organization". Like a cone-in-cone structure, where cones of varying diameters can be placed one within the other, a three-tier organizational frame is stipulated. It begins with the innermost core of the Gurukula organization which consists of "Guru who is the head of the institution and the fully dedicated first degree disciples". The second layer is the second degree disciples which form the peripheral aspect of Gurukula. It comprises those who might belong to the "Greater Gurukula or the Yellow Fellowship or other affiliated formations of the Gurukula movement". As we read this section further, we understand that from Guru, through the first and second degree disciples, the conceptual frame of the Gurukula is expanding to a universal dimension:

> Even this peripheral aspect of the Gurukula organization merges imperceptibly into the larger Gurukula movement which comprises the whole of the public without any limitation.

"Without any limitation" literally and figuratively put an end to the search for the organizational frame of reference of Narayana Gurukula. The findings began from the limited frames related to Indian context and finally reached the limitless. This limitless conceptual organizational frame merges with the philosophical frame, which is universal in nature. The philosophical frame is the ideals, objects and aims of Gurukula, which has been partly stated in the passage already quoted (paragraph seven), and specifically clarified in Section 2 of the memorandum:

> The primary object of the Narayana Gurukula Foundation, as has already been indicated above, is that of promoting or propagating irrespective of race, caste, nationality, sex or creed, etc. among the people at large that body of exact knowledge known as Advaita Vedānta (non-dual wisdom)

in its various forms indicated above (see Section 1). It attempts to preserve in its purest form the way of life of the original Guru Narayana for successive generations to emulate or follow.

Preserving the way of life of Guru Narayana in its purest form for successive generations is also specified as an aim of the Gurukula.

As corollaries to the primary aim, adding the principles of *ahiṁsā* (non-hurting of fellow beings) and the open universal and unitive outlook in life as implied in the motto "One Caste" (race), "One Religion" (faith), "One God" (supreme goal) completes the ideals, objects and aims of Narayana Gurukula.

So we have deciphered the different domains and dimensions of the frame of reference of Narayana Gurukula at the organizational and philosophical levels. Now let us reread the birthday resolutions of Nataraja Guru, given in the opening paragraph, in the light of this frame of reference. The five-point resolution was adopted in 1945. He went abroad in 1948, undertook a world tour for three years, returning to Varkala in 1951. By that time significant steps were taken to materialize all the five goals set by him on his fiftieth birthday. An account of how the resolution had been fulfilled was recorded by one of his disciples, Swami Chidambara Theertha, in page 248 of his book *Guruvum Vachanavum* (Guru and the Word).

The conceptual frame of reference of Narayana Gurukula is the "limitless" universe. The content of the birthday resolutions too were universal. Modern terminology allows us to interpret this universal frame and universal content, together with its fulfilment, as the vision, mission and action plan of Narayana Gurukula.

8

Founding Island Gurukula Aranya

Nancy Saumya Yeilding

As I contemplated writing this article to contribute to the celebration of 100 years of Narayana Gurukula, I experienced a semi-startled recognition that, for half of that span of time, I had been a part of the unfolding story. The element of "startle" has to do with how we tend to live in time, with the present occupying much of the stage, obscuring the reckoning of years. But "startle" also could be evoked by consideration of the earlier years of my life in the US, which in no apparent way would have led anyone to anticipate what was to come. And yet, the yearning to go beyond boundaries of social definitions and mental constructs gave rise to a variety of twists and turns in my life, until, like a turbulent, rushing river finding its home in the vastness of the sea, my searching led me to the feet of Guru Nitya Chaitanya Yati.

The mention of the feet of the *guru* evokes an ancient archetype, one that in today's world can seem archaic, and even laden with patriarchal overtones. But at its core, it is a precious reminder of humanity's wisdom heritage and an appropriate grateful reverence for and attention to a person capable of and committed to authentic sharing of the teachings that nurture life and love. This bipolarity was at the core of Nataraja Guru's inspiration to found a *gurukula*, a home and family dedicated to the verdant possibilities of living wisdom. Although I had only the vaguest idea of all this at the time I met Guru Nitya, he manifested it in his very way of being in the world. What I was very clear about from the first day of knowing him was that entering into the light-filled circle of his loving wisdom required no difficult decision. Rather, it came as the most natural flow of life.

In the early years of my study with Guru Nitya, at his request, I began editing Nataraja Guru's *Autobiography of an Absolutist,* which tells the story of the founding of Narayana Gurukula. When I first read his account of what inspired him and the dedication that guided him through what can only rightly be called "trials and tribulations", my heart was deeply touched. As I read his honest recounting of the mistakes he made, the various forms of suffering he endured, and how he persevered through it all, I felt both compassion and admiration. There was something else, too, in my response, which I could not readily name or describe, but was a deep stirring.

Although Nataraja Guru's *samādhi* had come just months before I met Guru Nitya in 1973, through his teaching and living example, both Nataraja Guru and Narayana Guru became my teachers as well. For the better part of eight years I had the incredible good fortune of living, travelling and studying with him, soaking deeply into the profound wisdom of the Narayana Gurukula *paramparā*, as well as into the world's many sources of precious spiritual and philosophical guidance.

But every once in a while, in ways that were sometimes gentle and sometimes sharply serious, Guru Nitya reminded me that this lifestyle would not continue indefinitely, ultimately because he would some day die, and – most likely much before then – I would be pushed out of the nest! The first of these reminders came as a sudden shock, while the last of them came as an anticipated question, posed quietly and firmly. Over time, I came to understand, with increasing clarity, that – generous as he was with his love and guidance, which would continue whether he was near or at a distance – he could not live my life for me, could not and would not make the decisions I needed to make for myself.

The final question came shortly before our planned departures from the UK to opposite points on the globe. Prior to that, I had travelled with him to most of the centres of Narayana Gurukula, from the headquarters at Varkala to his home base in Fernhill. I had found that life in the *gurukula*s offered an unusual and precious combination of stability and freedom: the opportunity to orient life around contemplation and the learning and teaching of wisdom, to expand the model of the family to the possibilities of universal connectedness, to nurture an active, fulfilling reverence for life, and to celebrate beauty and creativity.

As the time for my return to the US approached, I realized that I did not want to give up this way of life. So when Guru again asked, "What are your plans for your life?" I finally had an answer: "I'd like to start a *gurukula* in the US, if I have your blessing to do so." He did give me his blessing, but, when I boarded my flight to the US, it was also a flight into the unknown. I was happy and hopeful, and I was also virtually penniless, with no idea where or how my dreams and aspirations might come to be. But, there are times in life when the unseen hand of the Tao manifests in completely unanticipated ways, perhaps particularly when we open ourselves to the unknown.

Although I was headed to Oregon and California, I flew into Seattle, Washington, because that flight was the least expensive to the West Coast. I went to visit friends, Nancy and David Leedy, who lived on nearby Bainbridge Island. When I shared my plan to start a *gurukula*, they were very interested and asked many thoughtful questions. I described the appeal of the simple *gurukula* lifestyle, but most of all I spoke of how the wisdom teachings shared in the *gurukula* classes and exemplified by the *gurus*' brightened minds and hearts, transforming doubts and sorrows to clarity and joy.

Later, as I was leaving, Nancy said semi-jokingly, "You could come start your *gurukula* on our other property on the island". Picking up on her light, casual tone, I demurred. As I travelled south to see friends and family, I took my dream with me, sharing it along the way. But my lack of funds was a sobering factor; there was no obvious path from dream to a true beginning. Then, one afternoon, while speaking with my mother, I said: "I'd like to not just be in an apartment, to be in a place with some space, where people could gather for classes, but how could I afford such a place? I'd like to be in a rural area, but near a big city …." At this point the thought came: "Has someone already offered me this?" I called Nancy and said, "I know you were sort of joking, but can we talk seriously about my starting a *gurukula* on your property?"

Not knowing what to expect, I was amazed by what she and David responded. They had purchased the land as an investment, so they were in no hurry to sell it. I could come and start a *gurukula*. If it flourished, the *gurukula* would buy the land from them when they needed the money.

If not, they would be free to sell the land at that time. My heart buoyant, I started my return journey to Bainbridge Island.

Nancy and David, concerned that I had not even seen the place, took Polaroid photos and sent them to me. Receiving them, I was even more eager, feeling like the dilapidated cabin at the grassy edge of the young forest was calling me. I arrived in May of 1981. Over the next six months, in order to make the cottage useable, I had to become an electrician, a plumber and a carpenter! Those days of hard work were also days of wonder, as unexpected blessings continued to shower. Guru Nitya came to conduct a *homam* to give a whole-hearted inauguration to Island Gurukula Aranya. Friends and family came to help, from California, Oregon and even Australia. They brought with them tools, energy, goodwill and camaraderie. Nancy and David were steady helpers, also connecting me with more friends and resources.

There were also difficult and discouraging moments, but Nataraja Guru's steadfast example was a continuous guide. By October, the cottage was cozy, with a single room large enough for a small group to gather for classes. One evening we celebrated with a warm and joyful *gurukula*-warming. When I announced, "now classes can begin", several friends said they were ready to join. The next week – just six months after arriving – I started teaching Narayana Guru's *Ātmopadeśa Śatakam*!

The morning after the *gurukula*-warming, we held a groundbreaking ceremony in a nearby clearing for a building to house a press to print the English *Gurukulam* magazine. I found employment on the island. Over three years, on weekends, friends and students came together with ingenious ideas and hard work to craft a two-storey building, mostly with trees cut to make space for the building.

Guru Nitya again came to the US in the spring of 1984. Before he came to Bainbridge, I travelled with him to California and Portland. Several friends had purchased tickets to join him on a ferry trip to Alaska. But in Portland he proposed that we cash in our tickets and use the funds to buy a printing press. We all readily agreed and a small team was sent into Portland on a press-buying expedition. They returned, jubilant at having made a deal to purchase a large offset press and other printing equipment. As I listened to the description of the four-colour press – 8 ft by 8 ft by 10 ft of heavy metal – I must have turned white, and started shaking my head, knowing that, even if we could get it in our newly-finished building, it would surely crash right through the plywood floor.

Guru looked at me sternly, but with a twinkle in his eye, and said, "Oh ye of little faith!" – Indeed! – Within three weeks of our return to Island Gurukula Aranya, funds had been donated for a concrete foundation and the wood needed to double the size of the building, and friends arrived from around the country to camp in the woods and form a construction crew. The day before Guru flew to Australia, the press arrived on a huge trailer, to be winched into place. In 1985, Island Gurukula Aranya's first issue of *Gurukulam* magazine began almost twenty years of publication.

As I contemplated the concrete slab (very difficult to remove), I knew the time had come to purchase the land, and, with Guru's encouragement, we initiated that process. After some years, with the participation of many willing contributors, and the continuing touch of grace, the land was permanently dedicated. Every year, friends gather from around the US and beyond for the celebration of our *guru-pūjā*, which links us to our ancient wisdom heritage, as it also provides an opportunity to share the wisdom-blossoms that are flowering in our lives to grace each moment.

As in all the *gurukulas*, Island Gurukula Aranya is a place where compassionate wisdom guidance is freely offered those who come to visit and to live, whether for days, months, or years. Early in the new century, following the wafting winds of change, I began teaching via email, connecting friends around the globe in shared study of the works of Narayana Guru and other wisdom texts. Then, with the advent of new technology, global connections were enhanced by live video discussions and meditations.

The potency of the seeds planted by Nataraja Guru 100 years ago thus continues to sprout and grow in ever-fresh ways and places, capable of surprising us, inspiring us and blessing us as we face the challenges of life, nourished by Narayana Guru's profound compassion and deep wisdom, which enfold us and speak to us, wherever we are in the world, reminding us that we–all–are one family, one Self.

9

The Portland Gurukula

Scott Teitsworth

THE Portland Gurukula has been in existence for just over half of the age of the Narayana Gurukula, which was founded by Nataraja Guru in 1923. The Portland branch sprang into existence in 1971, and my wife Deborah Buchanan and I have tended it in our home since 1978.

Very near Narayana Guru's 116th birthday, in late August 1970, Deborah and I entered Portland, Oregon, on the West Coast of the US, travelling from the east. Unknown to us, Swami Nitya Chaitanya Yati, soon to become Guru of the Gurukula, arrived in the same city from the west. Thanks to a happy accident, we noticed an advert in the newspaper and attended his first class in America, on the *Bhagavadgītā.* We immediately became admirers, and shortly after, disciples.

Nitya was such a fantastic discovery! Finally, here was someone who actually knew what he was talking about, after a lifetime (19 years by then) of being surrounded by people who were equally or more confused than I was, blowing smoke, in more ways than one. And he was not dishing out simplistic platitudes, where I could just be amused and go about my business unaffected: his challenging assertions immeasurably augmented my thirst for knowledge and comprehension. Over the years, he gradually turned me away from my "neurological agitations" to semi-serene contemplation and heightened clarity.

I have written about those early *guru* blessings in *Gee, You Are You,* and *Flip Flop*, articles now posted on my website, and in my audio book *Coming Back to Ourselves,* so I would not rehash them here.

After the end of that first *Gītā* class, with one session for each of the eighteen chapters, Deborah joined Swami Nitya for a trip around the world to continue her discipleship, while I worked to save some money so I could attend his next Portland class. From Australia in July, he wrote me words that would come true beyond my wildest surmises:

> All these months you were so far and yet were very close to my silent thoughts and prayers. Your growth and unfoldment are of great importance to me for more than one reason. The flowers that bloom and wither away in the hidden bushes of obscurity secretly proclaim the inherent potential of this good earth to conceive and generate beauty and goodness. How much more would you be a promise as a man and not a flower, in human society and not in the bush, in the wake of a new age and not in obscurity? How much I wish your youthful blood will pulse with my inarticulate prayers, and my spiritual visions find a home in your thoughts.

In September 1971, Swami Nitya founded the Portland Gurukula, which became home for several of his lifelong friends and collaborators. The diary he kept of its activities forms Part 2 of his autobiography, *Love and Blessings*. Portland was his home base for most of the 1970s, between dozens of trips all over the globe to address eager audiences, but in

1978 he began to gravitate more back to India, and many of his disciples dispersed. Having been married by him in 1977, Deborah and I took over the Portland Gurukula, and began holding weekly classes in our home, starting with listening to tapes of that extraordinary *Gītā* class of 1970, now sadly lost.

I have been teaching the *Gītā* ever since, recently including email correspondence courses, and spent the years 2005 to 2015 writing a detailed commentary of every verse, based on Nitya and Nataraja Guru's incisive understanding. I dared to do this only because of all his major works are in English, Nitya's book on the *Gītā* contains only a hint of what he revealed in his classes, which was truly life-changing.

As I worked painstakingly through my *Gītā* exegesis, taking plenty of time to contemplate each verse and read multiple perspectives, I began to suspect that Nitya had left a door open for me to bring his vision to full fruition after his *samādhi*. I felt as if he was looking over my shoulder with a critical eye, and I should be as careful with my own thoughts as I had been with editing his books. There is always doubt that one is doing justice to a great theme. And then, toward the end of Nitya's massive, three-volume commentary on the *Bṛhadāraṇyaka Upaniṣad*, I ran across what struck me as his endorsement:

> In India, tradition allows a disciple to complement the writings of his *guru* by supplementing the *guru*'s writing with what the *guru* should have said but did not say. Second, the disciple has the permission of tradition to rephrase the argument of his *guru* if the latter's words do not properly serve to establish and arrive at the conclusion intended by the *guru*. – III.440

What a relief! It allowed me to relax, to stop feeling like an imposter. Nitya's words always led luminously to his conclusions, but I definitely felt I was recording what he meant to say but had not gotten around to.

The history of my extensive editorial career may be found in *A Moldy Yogi Retires,* posted on my website. Although I imagined I "retired" from working for the Gurukula in 2018, more than twenty years after Nitya's death, in 1999, the projects keep on coming, as if by magic. I have never lost enthusiasm for them, because the Gurukula writings are so rare and special, as Nitya noted in the 1980s:

> During the last thirty years the Gurukula has produced a literature of its own which stands as markedly different from anything thought of, written, or presented by anyone else. I cannot adequately thank Nataraja Guru for his wisdom in programming such a course of studies, which has been elaborated year after year. It is an ever-growing wisdom heritage that commenced with the publication of *Values* in 1955 in Bangalore and *Gurukulam* Magazine in Varkala shortly thereafter.

When he wrote this, most of Nitya's major English-language works were yet to take shape. Since then, our weekly Portland Gurukula class has studied most of them in depth, and elucidations via globally shared Class Notes are archived on my website. I consider Nitya's exposition of *Ātmopadeśa Śatakam: That Alone – The Core of Wisdom,* to be the most profoundly stimulating spiritual exploration in existence.

Speaking of *Values* magazine, the Portland Gurukula maintains a modest archive of memorabilia, including Nitya's near-complete collection of *Values,* one of his walking sticks, the meditation bell from his mendicant days, some of his art. Most of it is a digital archive of books and articles. A few of the decaying old tapes have been converted, but need the handling of an expert, which I am not. The archive serves as a place to donate items, as the Guru's friends approach the end of their lives. As there is no successor in the Portland Gurukula, the archives will have to be passed along from here before too many more years have gone by.

What has mattered to me most about the Gurukula version of Vedānta is its alignment with recent findings in neuroscience and psychology. Moreover, thanks to Narayana Guru's grounding in an all-encompassing mystical vision, a balanced mind is not to be restrained from engagement in current affairs. Nitya expressed this often, here in a letter to first World Citizen Garry Davis in 1977 (Nitya and Nataraja Guru were the official *gurus* for his World Service Authority) reprinted in *Love and Blessings:*

> We cannot sit around as passive witnesses of the doom that is steadily and systematically coming to a crucial point. The watchword that was given to us by Nataraja Guru was to take upon ourselves unlimited liability as the *satya-dharmīs*, the responsible custodians of the conscience of humankind. – 450

A favourite example of Narayana Guru's keen intelligence is recounted by Nitya in *Love and Blessings:*

> When Gandhi had visited Narayana Guru in Varkala long before, he had insisted on the natural differences between people and argued that caste was as much real as a tree having big and small or tender and withered leaves. Narayana Guru had replied that all leaves of the mango tree taste the same when chewed. The Mahatma had said that caste was good for the transmission of a craft or a trade from father to son, thus preserving its tradition. Narayana Guru exposed the shallowness of his argument by pointing out the decadence of every craft in India, from masons and carpenters to jewellers and sculptors, because there was no

> competition in any field. Nor was there any organized effort to transmit the modern techniques of silk weaving, metallurgy, architecture and the like to aspiring and efficient novices. These progressive stands of Narayana Guru created a tug-of-war in my mind with the Mahatma's conservative approach. – 90

Although I never met Nataraja Guru, he was so alive in Nitya's life that I have an intimate connection with him and his writings. I have compiled the many stories about him appearing in *Love and Blessings* into a separate document, and it is posted on Nitya's website: http://aranya.me/read.html .

Toward the end of Nataraja Guru's life, a splinter group of mostly Westerners developed, and after his death in 1973 it formed its own Nataraja Gurukula. Its history will likely be lost with the passing of this generation, yet at least one story deserves preservation:

A Tale of Nataraja Guru

by Patrick Misson

> Years later, I was staying at the Gurukula in Ootacamund, high in the Nilgiri Mountains of Tamil Nadu, where the green rolling hills and the chilly climate reminded one of England. The Gurukula there was a large study hall, with various bedrooms giving onto it and the Guru's room at the far end from the entrance. I lived in a small hut a little higher on the hillside. As we got up at five and were hard at it studying, etc. until lunchtime, everyone would retire for a couple of hours' siesta after midday. One afternoon I came down to the main building to see if it was teatime and saw a bunch of small children come haring out of the door and off down the hillside. When I went in I saw that a lot of books had been pulled from the shelves and flower vases overturned, etc. Guru

was standing there and he told me that he had woken up on hearing a noise and had come out and discovered some local kids trashing the place. He had told them off and had grabbed the cheekiest boy and given him a two-fingered slap on his hand to chastise him, and the kids had run off. Guru then said that, although he had only smacked the boy's hand lightly (he demonstrated – just enough to sting), he had done it because he had lost his temper, which was incorrect. We had our tea and then Guru assembled the dozen or so disciples, put on his coat, took his walking-stick and we processed down the hillside to the nearby village.

Now you must understand that Guru was 70-something years old; he was a very famous person, and held in awe by the local people as a great holy man. Also, the neighbouring village towards which we were heading was an untouchable settlement – these people were the lowest of the low in Hindu society and were forced to live in this filthy ghetto – their touch and their very presence were polluting to even the lowest-caste Hindu. So when we started processing down the one sordid street, the entire population came out of their houses and stared in silence and apprehension. Guru called someone and asked them if they knew where was the house of the little boy he had smacked. He went up to the house where the parents of the boy were standing. They had their hands joined in *namasté* and were half-bowing to Guru, obviously fearful of what would happen after their child had disturbed the great man. The little boy came out, cowering behind his parents. Guru said that he, Guru, had acted wrongly. Then he went down on his knees and performed the full prostration – the ultimate traditional Hindu gesture of debasement – that is, he lay flat on his face on the filthy ground and with his hands joined in supplication he touched the feet of the little boy and begged his forgiveness.

> Why was he my Guru? Because when I saw this kind of thing I knew that this was a real man; this was what human beings were put on this earth to be – and if I could not become like him, I would at least serve him for the rest of my life.

Many of us have dedicated our lives to one or more of the Narayana Gurukula *gurus*, because their greatness calls us, and we cannot resist.

The Portland Gurukula recently celebrated its fifty-year anniversary, gathering together a gratifying number of the early enthusiasts, now mostly in our 70s. Only a few of us consider ourselves "members" of the Gurukula – we are just ordinary people, and the books we study together are simply sources of wisdom and an opportunity for collective pondering of the mysteries of existence. As such, the Gurukula is a sustaining and elevating addition to our lives. We share Nitya's spirit, with which I give him the last word, from *Love and Blessings* (621):

> The entire universe is sitting in the classroom of the Divine, learning the lesson that no teacher is better than oneself and everyone has to take responsibility for their own physical health, spiritual illumination, and ultimate fulfilment. The schools, teachers, textbooks, syllabi, and curricula are all the external incidents in offering one's worship to oneself.

Links:

Nitya's own site: http://aranya.me/index.html

Scott Teitsworth: https://nitya-teachings.weebly.com/

Scott Teitsowrth email: tapovana@hevanet.com

10

Hundred Years of the Narayana Gurukula

Deborah Buchanan

My first experience with the Narayana Gurukula was not as a place or organization but as words of insight. These words were spoken with experience and assurance in 1970 in a class on the *Bhagavadgītā* taught by then Swami Nitya in Portland, Oregon. I did follow those words to a number of specific places. First was to the Gevaert family compound and Gurukula in Saint Martens-Latem, Belgium. The antique words from the *Gītā* and their interpretation by Nitya took me into the heart of old and new Europe, then on to India, north and south, to Gurukula compounds, to family homes, and into the still-wild remnants of the subcontinent's forests.

Vivid in my mind is the swept-sand compound in Varkala with its old, old mango tree providing shade for afternoons of quiet conversation and meditation, as well as the nearby hillside where Nataraja Guru stayed and offered early morning coffee classes on the *Saundaryalaharī* as dawn rose over the palm trees. The Gurukula in Ooty in the Nilgiris mountains, where I stayed many times among the tea plantations, the cold sky, and eucalyptus trees, holds a special place in actuality and in my heart. Philosophy, tropical air, the breath of high mountains, the touching relations with other students – all these are intermingled in my definition of the Narayana Gurukula.

My first description of the Narayana Gurukula came at this time from Nataraja Guru when he recounted his idea of a Gurukula as a swimming pool with young women sitting around eating biscuits and cream. Their children were playing nearby and would ask the mothers, "What about our fathers?" The mothers would say, "Oh, don't bother about that". The surprising surrealism of that image still makes me burst out laughing.

Over the years numerous places have been associated with the Gurukula – in Singapore and Malaysia, Australia, Europe, but most especially in the United States in Portland, Oregon, and in Bainbridge Island, Washington. From 1970 onward Portland is where Guru Nitya came to teach classes and interact with students, Guru Prasad coming with him in 1990. In 1981, the Island Gurukula Aranya was inaugurated in Bainbridge, and in both Portland and Bainbridge study has continued over numerous decades.

For almost twenty years, beginning in 1981, the Bainbridge centre wrote, collected, printed and distributed the English language *Gurukulam* magazine. When that ended, production moved to the Portland Gurukula for nine years, 2004 to 2014. As continuators of the English language *Values* stewarded by Swami John Spiers at Kaggalipura (1955 to 1974), these magazines form a unique literature of their own, covering topics of social and political interest as well as art and philosophy.

I list these many Gurukula locations to note how varied and active they have been. It is, however, important to emphasize that place is not a defining characteristic of the Narayana Gurukula. A location can provide a base for the Gurukula but it is not the Gurukula itself. The essential factor is a search for understanding and the answering dialogue of shared discovery and insight. It is a "place" of interaction – through written word, through silence, through the physical presence of teachers and students. This is what Nataraja Guru envisioned in 1923 when he inaugurated the first Gurukula – wisdom pursued through contemplation and through the interchange of *guru* and *śiṣya*. We should remember the meaning of the very words themselves: *nārāyaṇa* referring to the incarnation of Viṣṇu floating on the serpent of endless time and to the abode of man or humankind; *guru* being the source of light and remover of obstacles; and *kula* the home for all of this. There have not been requirements to participate in the Gurukula. There is not a profession of faith or loyalty. There are certain texts often studied and shared – *Ātmopadeśa Śatakam* and *Darśana Mālā* come especially to mind – but they are not obligatory. Each teacher has delved into various wisdom traditions and shared that journey – the *Gītā* and the Upaniṣads, the *Tao Te Ching*, Carl Jung's writings, St. John of the Cross, Basho's haiku and others have all been on the let us-explore-this list. Students themselves bring their own histories and interests to the mix.

Boundaries of nations, strictures of culture – neither of these are essential aspects of the Narayana Gurukula, which expresses certain historical experiences and personalities but is most truly an open community that provides each student a place where they can pursue their doubts and deepest questions. What is wisdom? What is it to be human? How can we participate in a world of loving justice?

One time a person put this question to Guru Nitya, "Who are you? What are you doing?" His response was, "I am part of the web". In those pre-Internet days, he was not referring to the cybersphere but to the unlimited interactions and connections of all beings. He was and is one part of that overarching web, as are we.

Narayana Guru walked quietly away from his Travancore village and from his feudal world's determined role for him. He did not fight, he simply followed his inner vision – so many revolutions followed from that subtle shift. Nataraja Guru studied science in India and then went on to Paris for further study at the Sorbonne, following which he taught in a Quaker school in Geneva. Guru Nitya travelled throughout Asia, Europe, the United States and Australia, teaching and writing, delineating the wisdom of the *ṛṣi*s through the languages of psychology and poetry and art. Guru Prasad continues this legacy of dedication and openness. The inner realization in each of these *guru*s has been of our deepest luminosity and of the unlimited connection we all share. It is this we carry with us.

In the winter of 1979-80 at the Ooty Gurukula, Guru Nitya one evening admonished all of us there, saying that we think we are individually choosing a path of altruism and dedication but in fact we are the result of Narayana Guru's many years of *tapas* (restraint and dedication). As Gurukula participants, we want to safeguard the insight that Narayana Guru had, neither mangle nor downplay nor alter it, but step into our new worlds and share this generous clarity with open hearts. Not for just another one hundred years but for each infinite moment.

(Such) meditation has the final result, not of solving mysteries but rather of making the sense of the mysterious live in the heart of man forever.

Guru Nitya Chaitanya Yati

11

What Narayana Gurukula Means to Me

Peter Oppenheimer

THE Narayana Gurukula has proved to be both a seminal and a centralizing presence in my life from the early age of 23 to the advanced age of 76, at the time of this writing.

In 1971, after a series of what I have come to think of as "Divine Mistakes", I stumbled onto and into the Narayana Gurukula at Varkala, Kerala in south India. I had recently graduated from Stanford University and was staring down a life that seemed to be preordained for me, not particularly of my choosing. The assumed path struck me as a life of laborious conformity, in which I would somehow get myself selected for some high-paying job, find, court and wed a wonderful wife, and make enough money to provide a comfortable, affluent lifestyle for said wife and an unspecified number of children, who would presumably qualify for some equally prestigious education and highly remunerative occupation in order to have and provide for kids of their own. Rinse and repeat, ad infinitum.

The jobs on offer struck me as more menial than meaningful and were mostly in the lucrative boom industries of corporate consumerism in an economy which was forged on the backs of indigenous people whose land and livelihoods had been ruthlessly stolen and very lives snuffed out, as well as on the scarred and bruised backs of an entire race of African Non-Americans who had been kidnapped and enslaved in order to produce the wealth and comfortable lifestyles of the colonial settler class of European Americans who benefitted through the propagation of such inhumane and, to this day still not accounted for, grave injustices.

As a certifiable hippie in 1971, I encountered Nataraja Guru, Swami Nitya Chaitanya Yati, Muni Narayana Prasad and others at the Narayana Gurukula at Varkala. My life has been an unfoldment initiated, guided and sustained by these three teachers and role-models, as well of course, as with their shared source of inspiration, Narayana Guru. The Narayana Gurukula has helped me to continue to cultivate the best of my hippie heart, along with teaching me a discipline with which to weather the storms of a life inevitably marked with failures and successes, pains and pleasures, and deep sorrows as well as prodigious moments of grace.

To understand how I got hooked by the Narayana Gurukula, it is helpful to understand where I was coming from. The hippy subculture arose as a reaction to the hypocrisy of preaching a "Love Your Brother" ethos from the pulpit on Sundays and a dog-eat-dog, economic ethos (e.g. pay him the lowest possible wages while charging him the highest possible prices) the rest of the week, and as a reaction to the conformity and drab uniformity of thought and behaviour, as exemplified by the grey-flannel-suited, necktie-wearing costume of the white, white-collar worker of the upper middle and upper classes in our society.

In a reaction to this, hippies celebrated an audacious freedom of self-expression, long hair on boys, short skirts on girls, colourful and unorthodox fashion choices, and enlarging gender and sexual orientation norms. A primary tenant of hippie philosophy as exemplified in the slogan, "Make love, not war", was the de-escalation of violent and militaristic approaches to conflict resolution. Hippies also spoke out and agitated for social justice and civil rights for marginalized and exploited people and against the self-destructive plundering and pollution of our natural environment, which is our source of life, sustenance and inspiration.

Back to 1971, when I was mistakenly and involuntarily dropped off at the gate of the Narayana Gurukula at Varkala (long story, recounted in my book, *Mirror by the Road*), I was feeling alone, misguided and somewhat bereft. I was looking for something other than what the Gurukula had on offer, which from what I could tell was a residential school teaching religion or philosophy of some kind. Having been turned off to religion by the aforementioned hypocrisy and to philosophy, which in my experience was an impotent ivory tower preoccupation of academics, removed from the rough and tumble of everyday life, I was not interested in either.

But Prasad, who then as now, was managing the Varkala Gurukula, responded to my announcement that I was looking for a place to learn *yoga* exercises, by saying, "Well we don't teach *haṭha-yoga* (physical training) here, but we do teach the *yoga* of the *Bhagavadgītā*. It's called *jñāna-yoga*". I have since come to learn that *yoga* explicitly refers to "union with the All". *Haṭha-yoga* is itself a relatively small subset of *karma-yoga*, which is union with the All through action. *Jñāna-yoga*, on the other hand, is union with the All, through wisdom, and *bhakti-yoga* is union with the

All through devotion. "In any case", Prasad continued, "you seem to be tired. You can come in and stay with us. When you have your strength back you are free to leave."

When I accepted his invitation, I was given my own cottage in which to sleep and hang out. It had a small patio on a hillside, overlooking a canopy of coconut palms and mango, jackfruit and papaya trees. From the patio I could hear the intoxicating laughter of children playing unseen down in the groves below.

I was told when and where the *āśrama* meals would be served and at what time and where they conducted "prayers". All the "inmates" (the irony of that word has become, for me more, not less, humorous during the past fifty years), as they referred to whomever was residing there, were cheerful and playful and seemed to want nothing from me, but only wanted to look after my well-being. This was all rather foreign to me and somehow at once both comforting and disorienting.

My second morning at the Gurukula, I saw another foreigner there, a hippie disciple of Nataraja Guru, the currently absent, yet obviously revered, founder and head of the Narayana Gurukula. This other foreigner was making use of the Gurukula's extensive library to research a paper he was writing. We talked, and at one point I asked him, "Hey, how much does it cost per night to stay here? What should I pay?" His immediate response took me by surprise, "Well, what Nataraja Guru says about that is, 'Just be as natural as possible'". I said something like, "What? You mean it might be most 'natural' to pay nothing". "Yes, it might be."

Now, that was something unheard of where I grew up: a free (or even by donation only) cottage in which to stay with a bed, bathroom, a desk with a chair and a deck with a view. I found this astounding and disarming. And it was at that point that I began to become keenly interested in the question: What is the philosophy or way of looking at life, of these people, which allows for this measure of cheerful altruism with benefits abounding to the world at large?

I asked Prasad that question, and he gave me Nataraja Guru's biography of his teacher, Narayana Guru, *Word of the Guru*. There I discovered enchanting stories of young Nanu's childhood and his precocious, steady ascent to the peaks of spiritual wisdom, beneficence and plausible sainthood. It was a great read, but at that point, I am afraid that most of the pearls of Advaita Vedāntic wisdom, tucked into those stories, sailed over my head or remained as seeds to sprout and blossom forth in later years.

Accordingly, after ten days or two weeks at the Gurukula, I was planning on leaving. I was off to rendezvous with a friend whom I had met up in Goa, as we planned to travel together to Ceylon (presently Sri Lanka). The day before I was leaving, likely never to return, Prasad approached me. He told me that the next month the Gurukula was hosting a World Parliament of Religions at a nearby lake site. There were going to be representatives from a number of world religions (Hindu, Christian, Muslim, etc.). He told me that they did not have any representation from China, and that he had noticed I had been walking around for the past ten days carrying the book *The I Ching*, an ancient Chinese Taoist wisdom text and mystical oracle. He asked if I would be willing to come back for that parliament and present a talk on *The I Ching*. Flattered and honoured, I agreed. And if not for that, the past fifty years of my life would have been dramatically different.

It was at that World Parliament of Religions that I met and attended classes with Nataraja Guru, as well as my getting to know Swami Nitya Chaitanya Yati, who would become my own *guru* and with whom I would live, travel and study for the next five years intimately, extensively, and almost without break, and then off and on for the following twenty-five years after that until his passing in 1999.

During these early days of my association with the Gurukula, I asked both Muni Prasad and Swami Nitya for the meaning of the word, *gurukula*. A common denominator in both of their responses was that the word referred to the *kula* of Guru, and that *kula* combines within itself a triple meaning. A *kula* combines, in a single setting, a home, a school and a temple (or place of worship).

A *gurukula* refers to a home in which the *guru* and others with him/her live as a family. It is a school in which the wisdom of the *guru* is taught, through words, gestures, behaviour, demeanour and silence. And a *gurukula* is a place for worship and devotion, in which the worship is not intended to be for an individual person, but rather something greater, which in this system of wisdom is referred to as the *Word of the Guru*, which can equally be heard in the buzzing of insects and the warbling of birds, and felt in a breeze or in the sweet thrill of receiving (or giving) affection, as well as every other aspect of our experience, once we have cultivated receptivity to this omnipresent glory.

A *gurukula*, being a common home dedicated to illuminating, interpreting, sharing and celebrating the Absolute, means that everyone else living, studying or worshipping at the *gurukula*, automatically become family to one another. And in the case of the Narayana Gurukula, I have found over decades this to be an actual dynamic, in that many of my dearest and closest "family" around the world are connected to me through the Narayana Gurukula.

In this regard, Guru Nitya enjoyed the saying of Jesus, "When two or more people are gathered in my name, I am there with them". And indeed being with my Gurukula family around the world always elicits that feeling of invocation, presence and grace. This is the magic of what Buddhists refer to as the *saṅgha* (congregation), which is one of the three treasures (along with the Buddha and the *dharma*) of Buddhism. In this regard I want to make special mention of the roles played by the two centres of Narayana Gurukula in Portland, Oregon and Bainbridge Island, Washington in nurturing this sense of familiarity and family among an ever-growing number of student/followers in the US.

A hallmark of the sense of family (*saṅgha*) of the Narayana Gurukula is not limited to those adopting any particular religion or philosophy. The Gurukula is explicitly dedicated to being open and welcoming to all, regardless of their religious or philosophical affiliation. And unlike in the case of many, if not most, religions, schools of thought or behavioural patterns, this openness is not offered with the intention to convert, but is offered in the spirit of acceptance and an eagerness to learn from the other and be enlarged by the very beliefs that may differ from one's own.

A *gurukula* is not just a home for an expansive family, but is also a school in which is taught the wisdom which illuminates, liberates and unites each with all. The Narayana Gurukula refers to this teaching as *brahma-vidyā* (Science of the Absolute). The Absolute includes all existence, all knowledge and all types and degrees of affection. Consciousness is seen as the solvent in which all seeming separation dissolves. Everything we can ever experience is experienced in and as consciousness. It is a consciousness which exists and which is dear to us. The nature, scope and dynamics of consciousness is both the object matter and the subject matter of the wisdom imparted by the Narayana Gurukula.

The Narayana Gurukula is devoted to the identification and promotion of what Guru Nitya often referred to as "the Good, the True and the Beautiful", all of which reflect that which we each and all are by our very birthright. That is the study and practice of Narayana Gurukula.

A hallmark of consciousness is non-duality, which is perhaps best described by the observation that while the sentence "I see a tree", can grammatically be divided into subject and object, in the experience of seeing a tree there is no such division. Nowhere can one demark, within such an experience, where the subject I ends and where the object tree begins. In the actual experience of "I see a tree", there is no such duality; the experience of seeing a tree is all of a single piece.

What is taught as wisdom at the Narayana Gurukula is not the dry syllogism of an intellectual exercise, but the full-bodied appreciation of all that life has on offer, as well as a deeper joy that consists in seeing the joy of others as your own joy. Such an expansive sense of Self and openness to all guarantees that, in equal measure, the sorrows and suffering of others will inevitably become more real to you as well. In any case, you will have a load of your own to deal with or deny. But, in truth, intimacy is its own reward. As is rhetorically asked in the *Īśāvāsya Upaniṣad*, for one who sees all others in one's own self and one's own self in all others, what delusion, what revulsion or what fear can there be for one with such a unitive vision?

And finally, coming back to the third meaning of *gurukula*, which is a place of worship. What is revealed in the wisdom teaching is so magnificent that it is cause for reverence, celebration and glorification. The object of worship here is not a particular god or gods, nor in fact does it require a belief in any god. Nataraja Guru was even fond of replacing the religious word God with the more scientifically acceptable term "chance". He said:

> If indeed it was simply chance that has caused the coming together of atoms and molecules in such a way as to create an orange, with its particular chemical profile, and if indeed it was also chance that led to my having a tongue with receptors in it which so perfectly complement those chemicals so as to create my experience of juicy sweetness, then I would call that "Chance" with a capital "C", and I would gladly bow my head in worship of such Chance.

One more very practical purpose of the Narayana Gurukula, which Guru Nitya often referred to, is that it should provide an opportunity for any person to step outside of the river of their own kārmic current of life, so to speak, and be provided with a safe and quiet place along the bank, to sit back as a witness and contemplate the flow of that river.

Over the past fifty years, the Narayana Gurukula has consistently provided that space of contemplation and repose for me, as it has for thousands of others. Though I have tried in three books, it is impossible for me to adequately gauge the beneficent impact this has had on my life, work, leisure, politics, social activism, connection with others, inner vision and sense of spiritual vitality.

Rather then, let me conclude with this feeling of immense gratitude and deep appreciation. Happy 100th Anniversary, Narayana Gurukula. And many more….

12

How Nataraja Guru Changed My Whole Life

Swamini Brigitte

IT WAS during April 1969 when Nataraja Guru was on his world tour, that I was fortunate enough to meet him and be part of the classes he gave in Lathem St. Martin in Belgium.

The main purpose of Guru's tour was to meet and converse with various professors and intellectuals from the Western world concerning his magnum opus *An Integrated Science of the Absolute*, which is based on the main philosophical work of Narayana Guru, *The Darśana-Mālā* (*A Garland of Visions of the Absolute*). In this, as also in his other writings, Nataraja Guru was breaking new ground by using a schema that is able to place correctly all philosophical disciplines into a structure. As Nataraja Guru himself says in his *An Integrated Science of the Absolute* in the fourth chapter, Māyā-Darśana, vol. 1, p. 531:

> We see from this how structuralism offers us an instrument wherein different disciplines claiming to be scientific can all be classified, compared and referred to a central structural norm.

By using what is called protolinguistics one is able to picture with the aid of a structure, the main positions of philosophical thought visually (i.e. perceptually) and then to be able to enter more easily mentally (i.e. conceptually) into the different disciplines, and so bringing these two aspects dialectically together to cancel them out against each other as counterparts, thus simplifying to a large extent that which is, to say the least, difficult to comprehend.

During that time, many of the youth in the Western world were in revolt against the established values which society embodied, they being primarily materialistic. Many young people were travelling and searching for an alternative way of life, with at least some spiritual values. Also, many young people were turning away from the universities and colleges, as the gap was widening more and more between the theoretical and living aspects of life that were taught in these institutions. Nataraja Guru arrived during this time with the answer, a teaching which had a direct bearing on both life and its inner values. Having finished his work on *An Integrated Science of the Absolute*, Guru was engaged in uncovering the hidden secrets of the *Saundaryalaharī* (*The Upsurging Billow of Beauty*), a work written by Śaṅkarācārya. Then even to Nataraja Guru's surprise, many people from all parts of the world spontaneously gathered in Lathem to attend his classes and imbibe his words of wisdom. Before Guru left Belgium seeing the enthusiasm and the interest shown for his teaching, he gave an open invitation to all concerned to come to India and participate in the World Conference for Peace through Unitive Understanding, to be held in November of that year at a place in Kerala near Kannur called Ezhimala, where a Gurukula called "Island Home" was being established.

How I Joined the Gurukula

I too was looking for a purposeful life being disillusioned with the values offered by society in general. So, in October 1969 I left Belgium and within a month of travelling overland by bus and train arrived in India. The very name of the conference, World Conference for Peace through Unitive Understanding, showed that it was not through action but primarily through knowledge (Unitive Understanding) that peace could be established. The conference started on Armistice Day (i.e. the date observed every year as the end of the First World War, which was on 11 November 1918). For eleven days people from all parts of the world came together and were given the opportunity to talk, listen and discuss on the various subjects open for discussion. All ate, lived and cooperated together, and in the evenings the local people also participated in the cultural programmes. After attending the conference, some stayed on to live under the Guru's guidance and attempted a new way of life.

What is this way of life that Nataraja Guru talked about wherever he travelled? In India it is called a *gurukula*, that is a wisdom centre for wisdom seekers; a neutral island; an association of like-minded people where teacher and taught live together; a way of life balanced between study and work; also aiming at self-sufficiency in a peaceful and quiet place. Here we now quote Narayana Guru's poem *Ashram* translated by Nataraja Guru.

Ashram

Narayana Guru

(Translated from Sanskrit)

Revealing the plan and intention of the original Guru.

In the Ashram proposed there should be a Guru
Learned, saintly and generous-minded
Who treats all with equanimity, calm dignified
Who is in control of his sense attractions.

Altruistically inclined and compassionate to the needy,
Who is truth-speaking and able,
Given to good conduct and prompt
In all matters necessary to do.

Having accepted the leadership of such a one,
Should be constituted an auspicious assembly,
Whoever comes to such a man
Shall all consider themselves as brothers.

As in this case even here, likewise
For women and for men, separate, for each
Should be created in different places
Schools as well as ashramas *always.*

For each of these there should be a head
Each one of them well qualified;
And all these treated together
Shall be known as advaita-ashrama

It was in this way that after the conference I was able to join the Gurukula and live in the Gurukula Island Home, where Nataraja Guru had put Swami Freddy in overall charge, and I was put in-charge of the women's section, a rare thing in those days, for there were hardly any openings or places for women to lead a spiritual life as such. So, from November 1969 till Guru's *mahāsamādhi* on 19 March 1973, I was very fortunate to live under the guidance of Guru. He taught and guided me during these years, especially instructing me on how a woman can live a full life within the Gurukula, functioning as a woman with all the qualities that belong to womanhood.

Before Guru's passing into *mahāsamādhi*, he earnestly requested us with these words, "Do everything you can to continue with the Gurukula". This was his dear wish which he imparted to his disciples, adding that the whole world should be full of small *gurukula*s. We are continuing with the Gurukula to the best of our understanding and ability to the present day, now for more than fifty years.

Guru Nitya Chaitanya Yati and Guru Muni Narayana Prasad, in that order, as the successors in the *paramparā* have been spreading the teaching, upholding the Gurukula and printing all the works of Nataraja Guru, which is perhaps the greatest service for the propagation of the teaching of Narayana Guru. Also, they have both expounded on the teaching with their own writings. The headship now rests in Guru Muni Narayana Prasad's dedicated and capable hands.

The Practical and Wisdom Aspects of the Gurukula

On entering the Gurukula, one's life changes and by that, over a period of time, one's thinking also changes from a more social and specific, to a more general and philosophical perspective. As the scriptures state, the willingness to serve is the first and foremost criterion. As quoted from Nataraja Guru's *Word of the Guru*, p. 115:

> Traditionally, then, the aspiring youth arrives at the abode of Guru who generally lives away from society in a forest hermitage. The future *śiṣya* (disciple) is expected to bring a token bundle of firewood as a sign of his willingness to submit to the household discipline required of him when he enters the *gurukula* (the family of Guru).

This antique picture shows the necessary attitude needed.

Some people are more intellectually disposed and some others more practically orientated but those who quickly assimilate intellectual ideas do not necessarily attain to the depth aspect of the teaching. Others of a slower disposition can attain, over a longer period of time, perhaps a deeper understanding. Theory and practice have to balance each other dynamically, with no hard and fast rules. Each person has to be guided individually. There are no short-cuts to wisdom.

A daily routine is a necessary base for an aspirant of wisdom. The balance between the physical and the mental aspect should be blended harmoniously for a happy and fulfilling life, which is in fact a *yoga* in itself. The balance is dynamic and changes with the different stages of life, and according to each person's disposition, but the focus should always remain on the *guru* and the teaching. As Ramana Maharshi states in his *Maharshi's Gospel* (Sri Ramanashram, Tiruvannamalai, South India 1946, p. 38):

> The *guru* is both "external" and "internal". From the "exterior" he gives a push to the mind to turn inward; from the "interior" he pulls the mind

> towards the Self and helps in the quieting of the mind. That is *guru-kṛpā* (grace of the *guru*). There is no difference between God, *guru* and the Self.

How can one cultivate and maintain this balance? This is where the *guru* is of the utmost importance, he is the catalyst. For, who can see oneself from an impartial standpoint and see the conditioning factors which make up one's social-relative value system, which is based on our beliefs and personal likes and dislikes. Even with the best of intentions, without being guided by universal principles, we are bound to be mistaken leading ourselves and others astray.

It is the *guru* who can guide us on the correct path to wisdom. So, in this way it is imperative that one has full faith in the *guru* and the teaching which he represents through his person and qualities, as stated by Narayana Guru in his poem "Ashram" as quoted above.

It is the willingness to listen to the words of the *guru*, that is most important and which most of the time will cut through the ideas and beliefs one has cultivated in earlier life. For example, most people think that existence is life, but one truly lives when one has a higher purpose than one's own I-sense. As the ego recedes happiness and peace wells up within oneself and after a short time in the Gurukula one can see the change in the aspirant, as the early rising and a well-balanced routine reorients the spirit and opens up a vista to a new way of life.

Many are the people who want to teach, but few are the ones who wish to learn, the saying "A little knowledge is a dangerous thing" is good to remember here on one's spiritual journey. As stated in Narayana Guru's *Ātmopadeśa Śatakam* (*One Hundred Verses of Self-instruction*), verse 12:

With skin, bone, refuse, and many an inner factor of evil end
Wielding these, lo! one ego looms: This which passes,
Is the other: That Self which grows to perfection,
Oh grant the boon that it does not the ego swell!

What I have shared with the reader in this article is only part of my ongoing journey in the Gurukula and covers only a small portion of the ideas and experiences I have gone through. It is by no means final or conclusive. We hope we have been able to share with the reader at least some of the precious values that the *guru*s and the Gurukula embody.

May the Gurukula live on forever!

13

Notes from the Neutral Centre

Swami Vyasa Prasad

THE basic questions of Vedānta, which concern every seeker, are presented in the opening verse of the *Brahmavidyā Pañcakam* (*The Science of the Absolute*). They are, *"ko aham* – this 'I' here what is it?" and *"kuto jagat* – whence this world phenomenal?". The latter question concerns the contemporary *Weltanschauung* (comprehensive world view) aptly described in the opening verse of *Advaita Dīpikā* (*Lamp of Non-duality*):

> *Names in their thousands, mental images in their thousands,*
> *And the corresponding inalienably emerging objects in their thousands –*
> *These together constitute the world.*
> *Just as dreams persist to seem real so long as the dreamer does not awaken;*
> *But when he awakens, the entire dream becomes merged into himself.*

Names, mental images and dreams have their basis in distinctive *zeitgeist*s which have defined each epoch: myths shaped the ancient world, religion dominated the medieval era, while science propelled modernism forward. Subsequently, liberalism introduced the post-modern-era, reason and faith intertwined in the post-post-modern, leading to the emergence of what the Argentinian-Mexican philosopher Enrique Dussel refers to as "transmodernism", accommodating both tradition and modernity. This cultural and philosophical movement, influenced by various philosophical currents fosters an inclusive dialogue between East and West. The nineteenth-century sociologist Gustave Le Bon defined modernity as the "age of crowds". With the advent of digital media the age of crowds is overtaken by the age of swarm. Crowds have a "we", the swarm does not develop a "we". The conscious personality of the individual in a crowd is submerged and the collective crowd mind dominates; crowd behaviour is unanimous, emotional and intellectually weak. Information has overrun thought in which the same algorithms are employed by Facebook, the stock market and the Intellect Services. Big Brother has been succeeded by Big Data.

Together with the passage of time, we have location in space. Euclidean space is geometric, following Euclid's axioms. Cartesian space is based on coordinates. Newtonian space, like Euclid, is infinite, homogeneous and completely featureless, one point being just like another. In 1905 the discovery of curved space introduced Einstein's conception in which space is no longer the stage on which the drama of physics is performed; it itself is one of the performers; for gravitation, which is a physical property, is entirely controlled by curvature, which is a geometrical property of space. Cyberspace is a notional environment, a virtual environment created by interconnected computer systems and networks in which communication over computer networks occurs. The Internet is a network of billions of computers and other electronic devices which makes it possible to share content very quickly across the globe. The World Wide Web enables non-specialists to share content via the Internet. Traditional space was organized around physical locations. However, in the Information Age, we have what the sociologist Manuel Castells Oliván terms "space of flows", where the key organizational principle is the flow of information, capital and power. It is a situation of decentralization, high-speed information flow and the transcending of national boundaries.

The networked globalized world is interconnected, but it is not "One World". It is fragmented and divided by leaders with precipitate thoughts, proclaiming their own opinion as the best, forgetting the most primary of human interests. It is impossible in the nature of things that only one opinion should prevail (*Word of the Guru*, p. 13). The methodology to resolve conflicts is geodialectics, based on the common interests of humanity. Fear and insecurity arise due to the lack of Self-knowledge. "When Self-knowledge shrinks, then ignorance is fearful" (*Darśana Mālā* I.7). The valorization of *vita activa* (the active person), sets in motion the dormant twin-roots of ignorance and craving. For wisdom to prevail, right thinking must precede right action.

Instead, thinking is burdened by increasing workload and an excess of stimuli and information which characterize late modern society. Consequently, contemplation, the cornerstone of culture worldwide, is stifled, leading to the erosion of inner peace, outer harmony and creative imagination. The contemporary Korean-German philosopher Byung-Chul Han describes the relentless striving for achievement and excessive positivity as the "burnout society". He criticizes multitasking, the ability to do two or more tasks simultaneously, saying:

> The attitude towards time and environment known as "multitasking" does not represent civilizational progress. ... Rather, such an aptitude

> amounts to regression. Multitasking is commonplace among wild animals. It is an attentive technique indispensable for survival in the wilderness.
>
> – *Burnout Society*, p. 12

Advancements in computer processing facilitate multitasking, whereas humans are enriched by the qualitative value of contemplation. Negativity is the dialectical counterpart of positivity and, equated together, they are the *yin–yang* of reality. There is a tragic vision of life which was an integral part of all old civilizations but is blotted out by the increasing glamour and fever of modern life. The Greeks celebrated their tragic playwrights Aeschylus, Sophocles and Euripides. In *Gītā* VIII.15, Kṛṣna refers to the world as *duḥkha ālayam aśāśvatam* (this transitory abode of suffering). Buddhism stresses that the true nature of all existence is suffering. A person should not get carried away by excessive positivity and overlook the negative pole of existence.

"It was a mistake to objectivize" (*Experiencing the Īśāvāsya Upaniṣad*, p. 1). The objective perspective came together in the seventeenth and eighteenth centuries under the heading of "classical science" or "Newtonianism", depicting a deterministic, clockwork universe. The scientific method liberated society from ancient superstitions and medieval irrational notions, but imposed a rigid deterministic model which did not have a place for randomness or chance. By the early nineteenth century, thermodynamics challenged the timelessness implied in the mechanistic image of the universe stating that entropy, the measure of disorder of a system, was increasing. However, Darwin's followers postulated that biological systems were becoming more and more organized. It seemed that the universe is simultaneously organizing and disorganizing itself! The Greeks understood *Logos* as the causal and primordial intelligence which is existentially continuing in and through creation as *an eternal creative and dissolving force.*

At the advent of the twentieth century Darwin, Freud and Einstein fundamentally changed the understanding of biology, psychology and cosmology. The famous Bergson–Einstein debate in 1922 concluded in favour of the scientist and the philosopher's perspective began to decline. The best European minds were drawn to be scientists. In his essay "The Barbarism of 'Specialization'" the Spanish philosopher Jose Ortega Y Gasset says that today there are "more scientists than ever, there are less 'cultured' men than, for example 1750" (*Great Essays in Science,* p. 128). When scientists become specialists they dislodge culture and their knowledge is confined only to their own little corner of investigation. It was a Faustian bargain in which higher subjective, synthetic and intuitive modes of thinking were replaced by a rational predictable clockwork model of the universe. The poet ceased to be a genuine seer. Instead, humanity began to rely on the scientist's work in the laboratory, which was reductionist and analytical, dissecting problems into smaller and smaller components, and which could at anytime be replaced by the paradigm shift of new theories. The Western world is particularly good at dissecting (divide and rule), building edifices with transient facets of the Truth, incapable of providing a synthetic vision of the Whole. In 1882, Nietzsche famously proclaimed the "death of God", marking a philosophical turning point. The eternal and timeless gave way to relentless progress leading to the horizontal expansion of technology into societies labelled "underdeveloped". The humanitarian and beneficial possibilities of science were eclipsed by the vested interests of corporations and the military industrial complex. The rational-intellectual-scientific mode of acquisitive thinking severed the contemplation of nature and the value of non-interfered serendipity. *Bhagavadgītā* IV.22 describes the *yogī* as *yadṛcchā-lābha-santuṣṭo* (satisfied by chance gains). The *saṁnyāsī* is *ayācaka* (one who does not ask or beg), content with what chance brings from day to day, and balanced between polarized attitudes of every kind such as fame or ignominy, etc.) (*Wisdom*: p. 250).Banished from classical science, Chance and Randomness resurrected with the advent of quantum theory. Nature gives us our good qualities and Chance sets them to work.

Technological advance under trained specialists dislodges culture and the decline of culture breeds the slow waning of society. Continuous technical improvement in the means of warfare occurs over the entire period of Hellenic and Roman decline. Even so, the breakneck speed of technological innovation today does not prevent societal decline. Modern technological prowess, devoid of normative principles, breeds a perilous arrogance, while material opulence, without cultural enrichment, leads to a profound sense of discontent, insecurity and unhappiness. Once regarded as the overarching canopy of all knowledge, philosophy has since fragmented into specialized branches, often prioritizing utilitarian values over pure theoretical pursuits. The modern outlook ignores the humanities in favour of science, technology and commerce. Metaphysics, ethics and aesthetics have been replaced by analytical subjects. The synthetic mode of understanding poetic symbols is muffled by literalism. In an expanding market economy, knowledge itself becomes commodified, copyrighted and exclusive. Without a common frame of reference, achieving integration of understanding becomes unattainable. Ritualism and commercialism creep into the domain of pure contemplation. Instead of certainty, quantum physics is pointing towards the cul-de-sac of probability and uncertainty. Nataraja Guru points out, "there is only possibility which gives certitude which impossibility alone can rule out. There is no question of probability" (*Autobiography*, p. 653).

Within the first decade of the new millennium, the documentary *An Inconvenient Truth* aroused public attention to the magnitude of environmental pollution, and the imminent climate and health crises resulting from contemporary consumption habits. Almost a century earlier, Narayana Guru had discerned the "senseless excessive greed of human beings". He said:

> The desires of animals are safely controlled by natural instincts from all abnormal excesses. ... The animals have no exaggerated needs like man. Man trots about the earth as a veritable demon of destruction. As he marches he carries behind him a trail of devastation. He cuts down the trees, and blasts and bleeds into paleness the green beauty of Nature for the sake of the plantations and smoky towns and factories which his unbridled desires necessitate. Not content with destruction on the surface he tampers with the crust of the earth, making it weaker and weaker day-by-day; and he covers the surface with miles and miles of iron and coal. Man is terribly inconsistent. – *Word of the Guru*, p. 12

Thinkers like Rousseau, Quesnay, Thoreau, John Ruskin and Tolstoy had similar misgivings about the global transformation from an agrarian society to machine manufacturing. Charles Dickens captures the mood in the opening lines of *A Tale of Two Cities:*

> It was the best of times, it was the worst of times, it was the age of wisdom, it was the age of foolishness, it was the epoch of belief, it was the epoch of incredulity, it was the season of light, it was the season of darkness, it was the spring of hope, it was the winter of despair.

In the midst of the first Industrial Revolution, the poet William

Wordsworth wrote:

Sweet is the lore which Nature brings;
Our meddling intellect
Mis-shapes the beauteous forms of things: —
We murder to dissect.

Enough of Science and of Art;
Close up those barren leaves;
Come forth, and bring with you a heart
That watches and receives. – *The Tables Turned*

The Gurukula has preserved the value of natural surroundings by reforesting native saplings, growing medicinal herbs, farming organically and managing waste sensibly.

Necessity is an inevitable and unavoidable aspect of life and the world of needs is to be kept in its place at the minimum so that attentive contemplative study is increased. Necessary aspects of life have to be willingly faced, while the contingent must be given free play in a spontaneous self-confidence (*Word of the Guru,* p. 82). The Roman philosopher Cicero said, "If you have a garden and a library, you have everything you need". The poet Wordsworth lamented, "Getting and spending we lay waste our powers". Mark Twain noted, "Civilization is a limitless multiplication of unnecessary necessities". A consumption-driven economy externalizes exploitation, masking harsh realities behind the deceptive facade of a contented life portrayed through sophisticated advertising tactics. An antidote is voluntary simplicity. While we rely heavily on machines for our daily functioning, the consciousness of a human being is distinct from the mechanical and repetitive nature of machines, and it transcends the roles of mere consumer, computer geek, or the capabilities of artificial intelligence. The "good life" is not cluttered with material possessions, but enriched with continuous contemplation on the Self (*ātmānusandhāna*).

While a posteriori science is based on empirical knowledge formulated in the laboratory, wisdom is largely a collection of axiomatic a priori verities and like the axioms and postulates of algebra and geometry, these are infallible in the certitude quality they contain. Myth is the deepest wisdom resource preserved in the collective unconscious with its archetypes (*BrUp*, vol. 1, p. 29). A healthy sober-mindedness is called for in the present age when norms must spell one world understanding (*Autobiography*, p. 622). The harmonization of various streams of thought into an inclusive symphony of values is possible with a wisdom frame of reference, dialectical methodology and Unitive Understanding.

If we should now examine the first question, "*ko aham* – Who am I?" we find that the query is of a perennial nature. The *Brahmavidyā Pañcakam* declares, *tvam hi brahma* (You are *Brahman* indeed). The *upadeśa vākya* (dictum of instruction) given to Śvetaketu in the *Chāndogya Upaniṣad* is *tat tvam asi* (That thou art). The Delphic maxim inscribed in the forecourt of the Temple of Apollo was *gnothi seauton* (Know thyself) made famous later by Socrates. Protagoras held that "Man is the measure of all thing". The poet Pope declared, "the proper study of mankind is Man". Guru Nitya writes:

> I am a questioning, recalling, figure-structuring, ensemble-interpreting, experience-evaluating consciousness, which becomes elated or sick through its own ability to project divine or diabolical phantoms that pass for the reality of the here and now. – Yati, *Narayana Guru*, p. 30

A section of of the Janaka–Yājñavalkya dialogue reads:

> If you see clearly that nothing of your environment is part of your being, then you will come to the conclusion, "I am the Absolute. I am pure". I have such clarity that the world is only reflecting in me and never touches my crystal-clear Self. Then you are free and you are inching towards perfection every moment.
>
> Environmental factors cling like dust on linen and patch it with Rorschach-blot type figures, and meddle with one's thoughts, ideas and beliefs. – *BṛUp*, vol. 2, p. 476

In the 1960s Marshal McLuhan argued that the medium through which we choose to communicate holds as much, if not more, value than the message itself. The emergence of signs and symbols, coupled later with advancements in technology, significantly influenced the dissemination of word wisdom. Narayana Guru imparted his final teachings by maintaining a profound neutral silence (*Word of the Guru*, p. 54). The absence of any medium ensured absolutism. Centuries earlier, the Buddha remained silent when asked certain metaphysical questions.

> The silence of the Buddha has traditionally been interpreted as profoundly significant. Correctly understood, it constitutes a seminal anticipation of the Mādhyamika, in both meaning and method. It was expressive of a profound insight of the Buddha, and its subtle implications were to be fully elucidated in the writings of Nāgārjuna. Conversely, the Mādhyamika is to be understood as the exploration and systematic expression of the Buddha's silence.
>
> – *Introduction to the Middle Way*, p. 6; see also *Word of the Guru*, p. 195.

This slight digression is of interest because of the relatedness between Nāgārjuna and Gauḍapāda, the author of the *Māṇḍūkya-Kārikā*, to Śaṅkara's composition *Dakṣiṇāmūrti Stotram*, which says *maunavyākhyā prakaṭita parabrahmatatvaṁ* (the knowledge of *Brahman* taught through silence), to Narayana Guru in the present age, whose meeting with Ramana Maharshi was marked by silence.

For centuries the sacred word was passed from generation to generation by oral traditions. This required each syllable be pronounced flawlessly. Following the development of the alphabet, the spoken word was painstakingly recorded on papyrus, palm leaf, parchment or silk. Śaṅkara's *Saundarya Laharī* depicts the Goddess holding a book (*pustaka karām*) (verse 15) symbolizing the reverence for written texts. Books were laboriously hand-written by scribes to be read by only a select elite. Writing also introduced the possibility of interpolation within a text. The advent of the printing press democratized access to literature, revolutionizing the spread of ideas. Subsequent breakthroughs, such as electromagnetic waves enabling radio and television, accelerated the formation of a collective consciousness. But this also enabled the spread of propaganda. Vocal people often overpower thoughtful and sensitive individuals. Today, the Internet and digital social media have exponentially expanded the collective consciousness, inundating humanity with a deluge of information and misinformation. Navigating this sea of content presents the challenge of discerning between sensationalism and genuine insight.

> Modern thought has created a type of *horizontal* explanation, in terms of equivalent effects and causes, all situated on the same level of empirical data, and this way of thinking has replaced the kind of *vertical* explanation that prevailed at an earlier time, and that linked visible effects to transcendent causes. – Robert Lenoble 1997: 19

In the *Chāndogya Upanisad* 4.4–4.9, Satyakāma Jābāla is instructed by a bull, the fire, a *haṁsa* (swan) and a *madgu* (diver bird). The final instruction, however, is given by the sage Gautama, his Guru, confirming the need for an authoritative and genuine teacher.

The twenty-first century has ushered a resurgence of religion and an interest in unifying science and spirituality. The foundation was laid by twentieth-century scientists. In 1927, Heisenberg introduced the Uncertainty Principle. Eddington spoke of subjectivism, structuralism and selectionism. Scientists published books sharing their insights with the reading public. Erwin Schrödinger's 1961 book *My View of the World* included a Vedāntic vision. Fritjof Capra's book *The Tao of Physics* (1975) explored the parallels between modern physics and eastern mysticism. The 1985 publication of Ilya Prigogine's *Order Out of Chaos* introduced spontaneity and self-organization in the universe. In 1990, Menas Kafatos and Robert Nadeau published *The Conscious Universe.* The 1993 publication of Amit Goswami's book *The Self-Aware Universe* shattered the Western scientist's belief that matter is the primary "stuff" of creation. It is consciousness that is the true foundation of all we know and perceive. The seventeenth-century philosopher Spinoza had a similar notion of "thinking substance" which formed the common basis of God, human beings and the world. Verse 79 of *Ātmopadeśa Śatakam* says, *cit prabhāvam ellām* (all is flux and becoming within the mind-stuff pure). The quantum discovery of non-locality as a fact of nature subverted the notion that the world is composed of individual objects. Today, neuroscientists like Donald Hoffman propose a bold theory – that objects do not exist independently of us perceiving them and that all that really exists is conscious agents. Anil Seth, another neuroscientist, emerging out of his materialistic background, says that the brain hallucinates its conscious reality. The Information Integration Theory of neuroscientists recognizes the fundamental subjective nature of consciousness, not a epiphenomena as a byproduct of brain activity. Christof Koch, a neurophysiologist and computational neuroscientist known for his work on the neural basis of consciousness, describes his ineffable experience of consciousness.

Science, in its search for the holy Grail of Unification, is moving towards the reality of consciousness, but lacks the normative notion of the Absolute. The integration of all physical and metaphysical disciplines can be accomplished through the principle of the Absolute as a Norm (*Integrated Science,* vol. 1, p. 129). Traditional Vedānta, nurtured on the compositions of Śaṅkara and his disciples, brilliant in itself, is unable to bridge the chasm arising from the insights of scientific research. An epistemological and methodological revaluation is required which is provided by Narayana Guru's compositions. The notion of the *karu,* the ontological essence of the physical and the metaphysical, is, in the words of the poet Novalis, "The seat of the soul where the inner world and the outer world meet. Where they overlap, it is in every point of the overlap". The subjective consciousness of the world within and the objective perception of the physical world outside are both treated as dependent factors and are the simultaneous manifestations of the *kāru* (mould). Both subject and object are simultaneous manifestations of the *kāru.* Reality can at once be the perennial projective cause that shines as knowledge within and as the changing panorama of the physical world without (*Narayana Guru,* pp. 10-11). In all dialectical counterparts the polarized opposites have a simultaneity in their origin and a binary interaction between them (*BṛUp,* vol. 1, p. 25). The methodological revaluation is seen in verse 10 of the *Ātmopadeśa Śatakam,* depicting two people sitting in a dark room, comparable to a controlled laboratory experiment. Thus, Vedānta needs revaluation to integrate and benefit from scientific knowledge, which better answers the basic question, *kuto jagat* (Whence this world?). The telescope and microscope have unveiled the marvels of both the vast macrocosm and the intricate microcosm. Space exploration has provided insights into the universe beyond the wildest imaginations of pioneers like Kepler and Galileo, while genetic engineering has facilitated the development of life-saving vaccines.

Where is happiness? Where is rest from the fever of life? Where is the image of perfection? Where is the fountain-source of wisdom from which a thirsty traveller can drink? Where is that luminous something, in which we can live apart and be free – free from a sense of want and suffering? (*Word of the Guru,* p. 3)

Vita contemplativa (the contemplative), who refuses to lie or use force, follows a negative rather than a positive form of teaching and a weaning process so as to avoid error in thought and thereby error in action. By paying attention to the problems of the individual, all conflicts can be resolved. The authoritative Word of the Guru is the compass that saves humanity from aimlessly drifting in the ocean of phenomenality. The Word, like Ariadne's thread, leads a seeker out of the labyrinth of *saṁsāra*. Narayana Guru's Universal Prayer (*Daiva Daśakam*) commences by addressing God as the navigator who steers the devotee out of the ocean of becoming.

Verse II.10 of *Darśana Mālā* says, *sarvam hi saccidānandam* (All is indeed existence, consciousness and pure happiness). Even animals differentiate a kind person from an unkind one. The collective consciousness of humanity is the repository of perennial wisdom. "All facts or events emerge from the homogeneous transparent fluid matrix of thought-stuff"

(*Unitive Philosophy,* pp. 177-78). The intuitive eye is able to penetrate the veil of the ever-transforming phenomena and imperience universal values and lasting happiness. It is not a mere fancy, but a well-grounded Science of the Absolute, available to each and everyone, irrespective of their background. In fact, wisdom is the underlying fabric of humanity. Nobody is denied entry into the palace of wisdom. All that is required is to shed one's false identification and conditioning, abandon any personal axe to grind, and open one's contemplative inner eye to perceive the numinous core which shines within and without. Feel the great wonder of your own consciousness that is continuously modulating as the world outside and the idea inside (*That Alone,* p. 5). One should adopt the attitude of Absolute Neutrality to Truth without any one-sided exaggeration or distortion, called in Sanskrit *yathā-tathyam* (as-itself-ness) (*Autobiography,* p. 622).

Protolanguage, quaternion fourfold structure, dialectical methodology and unitive understanding are terms in Nataraja Guru's lexicon. "Revaluation" and "restatement" refer to the methodology used by all wisdom teachers, and specially by the Gurukula *paramparā*, giving continuity and relevance to humanity's precious wisdom heritage. In this way wisdom is taken forward, breaking new epistemological and methodological ground. If such a way is not popular, it is, as John Locke said, "... without any other reason but because they are not already common" (*An Essay Concerning Human Understanding*). Nataraja Guru himself noted:

> ... these views of mine stood isolated because of the new lines of epistemology, methodology and axiology that I had developed with some bold originality derived from the Guru wisdom of ancient Upaniṣadic lore. – *Autobiography,* p. 648

Commentaries with scientific terms seamlessly correlated with Yoga and Vedānta nomenclature integrate the two streams of understanding, one obtained from the laboratory a posteriori, and the other from traditional wisdom texts, a priori, to arrive at unitive understanding, especially the nature of the mind – the functioning of consciousness in the individual – as a stepping stone to Self-realization.

It is therefore imperative that perennial wisdom be available in the contemporary idiom so that the Guru's "revolution in understanding" is widely appreciated. *Lingua mystica* of traditional texts is to be translated into *lingua scientifique,* understandable by the modern mind. In his commentary on *Bṛhadāraṇyaka Upaniṣad, adhyāya* V, *brāhmaṇa* 13, *mantra* 2, Guru Nitya writes:

> The teachings of the Upaniṣads need to be retranslated into the language of the newly educated so that these *mantra*s won't look like the lisping of a baby. ... So we solicit the favour of the reader to cooperate with us in our effort to correlate the ancient in the scriptures with the modern positive sciences. – *BṛUp,* vol. III, pp. 141-42

While imitation is easy, it takes courage and a genuine love for the welfare of humanity to delve into the foundation of Truth, refresh the epistemology and correct the methodology regarding fundamental principles. In the Introduction to *One Hundred Verses of Self-Instruction,* the commentary on the philosophical masterpiece, *Ātmopadeśa Śatakam,* Nataraja Guru describes a new world outlook, which is neither Eastern nor Western, neither ancient nor modern, neither academic nor religious, neither pragmatic nor sentimental. Let the Guru be praised for such an open and dynamic outlook (p. 9).

The front matter of the collection of articles titled *Unitive Understanding* is

> Dedicated to all kind and generous men and women, young or old, anywhere, anytime, who seek in open and dynamic terms to abolish and absorb all polyvalent rivalries of relativistic values into the real and crystalline fourfold structure of universal Selfhood that includes all opposites which is none other than that value of all values the Absolute.

The Narayana Gurukula is a haven for truth seekers, lovers of humanity, anyone eager to breathe the atmosphere of freedom in living terms, those who abhor violence in any form, and care for the welfare of all. It is a neutral space that abides in the measurelessness, timelessness, pure duration of eternity. Having established such a foundation, the task before its members is to neutralize those tendencies which promote divisiveness arising out of ignorance, engender people endowed with a philosophical eye, trained and enabled to speak without ignorance, prejudice or bigotry. Guru wisdom is actualized by its application in the here and now.

In conclusion, below is an excerpt from the very first issue of *Values*, titled "Islands of Neutrality".

The Gurukula movement is wide open to all who can appreciate its principles. As another experiment, it may be considered as merely writing upon the water or a sign on the wind-blown sands. But it calls attention again to the perennial message of all who have won through to free perception of lasting values. Its importance also depends upon the renewed enthusiasm of its inmates and upon keeping the living indicator at the neutral zero mark, neither deflecting towards escapist negation, nor towards mechanical positive success. Both ends have their tragic results, apathy or pride in attainments, the former largely in India, the latter in the orthodox religious and political ideologies of Europe and America.

Contemplation needs restatement, and contemplatives need a home. The whole world, including the people of India and the countries of the West, has need for wise persons, genuine philosophers, as guides and helpers. Right thinking precedes right action, but the people of the rushing modern world exhaust themselves too much to be able to begin to contemplate. The secret is lost for them.

In the Gurukula Movement it is fostered and is a living *raison d'être*. The ancient treatises speak out and divulge their great secrets when read by the dedicated seeker who is free from specialized backgrounds. The words of the Gurus become meaningful. The word of the living Guru is the spark to the tinder of understanding. Instead of a destructive atomic reaction, there is a chain-reaction of wonder and illumination.

Thus happiness prevails within these islands of neutrality set in the troubled sea of restless humanity. All can catch this Gurukula spirit. This is no private, exclusive, esoteric, copyright idea. There can be more and more islands in humanity's stormy sea, sanctuaries for lovers of truth and wisdom (*Values,* December 1955, vol. 1, no. 1, p. 26).

Nataraja Guru envisioned the Institute of the Science of the Absolute and formulated the Science of Dialectics so that wisdom is established as a high Value in life, ensuring the Happiness of all.

|| *auṁ tadeva sat* ||

References

Dickens, Charles, 1859, *A Tale of Two Cities*, London: Chapman and Hall.

Gardner, Martin (ed.), 1997, *Great Essays in Science,* Oxford: Oxford University Press.

Hallyn, Fernand, 1997, *The Poetic Structure of the World, Copernicus and Kepler,* tr. Donald M. Leslie, New York: Zone Books.

Han, Byung-Chul, 2015, *The Burnout Society*, tr. Erik Butler, Stanford: Stanford University Press.

Kafatos, Menos and Robert Nadeau, 1990, *The Conscious Universe: Part and Whole in Modern Physical Theory,* Fairfax, VA: Springer-Verlag.

Lenoble, Robert, 1997, *Poetic Structure of the World*, New York: Zone Books.

Nataraja Guru, 2003, *Word of the Guru*, New Delhi: D.K. Printworld.

———, 2006, *One Hundred Verses of Self-Instruction*, Varkala: Narayana Gurukula.

———, 2011, *Autobiography of an Absolutist*, New Delhi: D.K. Printworld.

———, 2012, *Unitive Philosophy*, New Delhi: D.K. Printworld.

———, 2012a, *Wisdom: The Absolute is Adorable,* New Delhi: D.K. Printworld.

———, 2020, *Integrated Science of the Absolute*, New Delhi: D.K. Printworld.

Natarajan, P., 1955, *Values*, vol 1, no. 1, Kaggalipura PO: Gurukula Publishing House.

Padmakara, Translation Group, 2002, *Introduction to the Middle Way,* Boulder, CO: Shambhala Publications, Inc.

Prasad, Swami Muni Narayana, 2010, *Shorter Philosophical Poems of Narayana Guru*, New Delihi: D.K. Printworld.

Wordsworth, William, Poetry Foundation: https://www.poetryfoundation.org

Yati, Nitya Chaitanya,1992, *Experiencing the Īśāvāsya Upaniṣad,* Fernhill/Bainbridge Island: East-West University of Unitive Sciences.

———,1995, *Bṛhadāraṇyaka Upaniṣad* (vols.I & II), New Delhi: D.K.Printworld.

———, 2007, *Narayana Guru*, New Delhi: Indian Council of Philosophical Research.

———, 2012, *Bhagavad Gītā*, New Delhi: D.K. Printworld.

———, 2012, *Psychology of Darśanamālā*, New Delhi: D.K. Printworld.

———, 2012a, *That Alone*, New Delhi: D.K. Printworld.

———, *Saundarya Laharī.*

14

An Integrated Science of the Absolute

The Science of Sciences

S. Omana

NATARAJA Guru (1895–1973) was the foremost disciple of Narayana Guru, and the founder of the Narayana Gurukula. He was a great scientist-philosopher who fully imbibed the vision of Narayana Guru, the great *ṛṣi* of the modern age. Nataraja Guru presented his Guru's vision in a way which could be understood both spiritually and scientifically. By comparing Guru's vision with the philosophical thoughts of the West, he emphasized its integrity and scientific nature. He renewed the value of the Guru's vision by assimilating it to the needs of the present world and making it acceptable to the present generation. Thus, the radiance of Narayana Guru's words spread throughout the world.

Among the books he wrote, *An Integrated Science of the Absolute* is his magnum opus. Nataraja Guru spent a whole lifetime formulating a unitive science wherein all disciplines concerning fundamental human questions could find a common ground. An integration where modern science and ancient spiritual wisdom could meet and merge like two opposite poles of a magnet. As a direct disciple of Narayana Guru he found the common ground in the ancient philosophical and spiritual discipline of *brahma-vidyā* which he calls the Science of the Absolute. He found that the groundwork for this Science of Sciences had already been done by Narayana Guru in his Sanskrit composition, *Darśana Mālā*. Nataraja Guru undertook the task to elaborate and represent it in terms understandable to the modern mind, particularly of the West.

Nataraja Guru's shorter works such as *Towards a One World Economics, World Education Manifesto, World Government Memorandum* and *In Search for a Common Language for Sciences* (all published in *Experiencing One-World*, D.K. Printworld, New Delhi, 1998) clarify how Integrated Science can be put into practice in human affairs with a global outlook.

Not only this world but all life is one. Science must also be seen as one, for only unitive science can guide human affairs in this Age of Science. Only such a vision, sustained by efforts to bring this vision home through a properly conceived education process, will ensure sustained peace in the world (Nataraja Guru 2001: ix).

The attempt to present the whole range of wisdom as a unified body of knowledge is not only evident in the East but also noticed in Western works, such as in Plato's *The Republic*, Aristotle's *Organon*, Francis Bacon's *Novum Organum,* and Fichte's writings. Such attempts, however, are mostly unsatisfactory or inadequate from a fully scientific point of view. A unified science based on the notion of the Absolute alone can serve as a regulative factor to effect orderly integration of all branches of human enquiry (Nataraja Guru 2001: 140).

The unified theory of Albert Einstein (1876–1955) reduces all the elements, categories or factors of reality, whether material or mental, to one homogeneous space–time continuum. This matrix, mental or material, conforms to the requirements of a neutral monism that has become acceptable even to positivists or empiricists of the present day. According to Erwin Schrödinger (1887–1961), matter and mind could be thought of as belonging together in the neutral zone or stratum of absolute Consciousness. Arthur Eddington (1884–1944), one of the philosophers of science, has made an earnest attempt to bring about a drastic epistemological revision, but he died before fully establishing his case. Henry Bergson (1859–1941) avowedly was an intuitionist but talked the language of pure creative movement visible in natural evolution of life. He modified and corrected the previous philosophers from the time of Plato. The thoughts of all these philosophers are considered to be vitiated in various degrees by static mechanistic or over-conceptualized abstractions tending to be only mathematical and estranged in various degrees from the living breath of reality.

Science itself in its most modern developments tends to support wholesale wisdom, rather than piecemeal trial and error methods annexing small territories of truth into its body. Science is undergoing a drastic epistemological and methodological revision. It is tending to include into its corporate body branches of knowledge like ethics and aesthetics which depend upon judgements of value. In doing this, it is possible for science to attain the status of a body of literature to be included alongside the wisdom literature of the world. When this is done we will have a complete Science of the Absolute. Such a science will have an epistemology, methodology and axiology of its own treating of existential, subsistential and value factors, at once. The question of scepticism or belief will not arise at all when Unified Science becomes fully normalized (Omana 2022: 234).

The Science of the Absolute can be also called a Science of Sciences, a Unified Science or an integrated body of knowledge. When such an enquiry is made so as to yield a common notion serving sciences, we then have a fully integrated Science of the Absolute.

One Consciousness underlies all branches of studies which for convenience sake, we call "sciences". If there is a science that concerns itself with this all underlying Consciousness, then that science should be the foundation of all sciences as well. Vedānta, particularly as visualized by Narayana Guru, is nothing other than that science. It is thus the Science of all Sciences (Muni Narayana Prasad 2003: 78).

How *brahma-vidyā* naturally becomes the Science of all Sciences (*sarvavidyā pratiṣṭhā*) is found in the *Muṇḍaka Upaniṣad*:

> *Brahman*, the creator of all and the sustainer of the world, emerged as the first of gods. He taught the science of *Brahman* (*brahma-vidyā*), the basis of all sciences to Atharvān, his eldest son. – Muni Narayana Prasad 1998: 6

Consciousness growing and branching out as all sciences (represented symbolically by the verses of the Vedas) is compared in the *Bhagavadgītā* to a (holy) fig tree (*aśvattha*) that grows upside down, rooted in the unknown high and branching downward forming the ever-growing world of human disciplines. Certain downward-growing branches tend to seek to know the root, they thus bend upward in their growth. Likewise, some roots grow down to become objects of study in the world here (Nitya Chaitanya Yati 1993: 329).

This is also implicit in Narayana Guru's restatement of Vedāntic wisdom, particularly in his *Darśana Mālā* (*A Garland of Visions*). In this work, Narayana Guru has been able to string together all the points natural or possible in the context of human understanding treated as a whole, a veritable garland of visions of the Absolute as the title claims it to be.

Nataraja Guru elaborated and reintroducesd *Darśana Mālā* as forming the nucleus of a Science of Sciences in his *Integrated Science of the Absolute. Brahman* being the one Absolute Reality that assumes the form of everything perceivable and conceivable in all the worlds, which form the object matter of all the sciences, a science that expounds that *Brahman* has necessarily to be the one Science that could find expression in all the sciences (Omana 2022: 235).

In this book, Nataraja Guru stated clearly how all branches of every field of discipline is a particular way for the function of the human mind to find expression.

Darśana Mālā consists of 100 verses of ten chapters with ten verses in each: Adhyāropa Darśanam, Apavāda Darśanam, Asatya Darśanam, Māyā Darśanam, Bhāna Darśanam, Karma Darśanam, Jñāna Darśanam, Bhakti Darśanam, Yoga Darśanam and Nirvāṇa Darśanam. These visions are structurally string together like precious stones forming a garland meant to be an ornament enhancing the dignity of man through wisdom. The garland of visions is the product of both inner experience and confirmation from outer textual sources.

When *Darśana Mālā* is thought of as a whole it has such a self-consistency at every stage that the question of proof does not arise. Each *Darśana* results from the meeting of the a priori and the a posteriori, wherein outer experiment neutralizes inner experience and as such yields both dialectical and as well as apodictic certitude at each stage.

When the garland is completely thought of, the reader will be able to recognize how the strings from opposite sides meet at the top or behind at the back of the wearer. At the most central part of *Darśana Mālā* (end of the 50th verse) is found one of the most famous of the great *mahāvākya*s of the *Chāndogya Upaniṣad, sat-eva-tat*.

This hangs as a secret pendent representing an extra jewel referring to the overall unity of the garland. The maximum richness of certitude is meant by Guru to reside at this central position. The two opposite ends of the string have to be tied together to make the complete epistemological, methodological and axiological circle. Within the scope of this circle, Narayana Guru has admirably succeeded in giving a symmetrically conceived equidistant position to all modern philosophical views, Eastern and Western, and also all significant modern scientific knowledge so as to present a sufficiently attractive edifice representing a wisdom free from geographical and traditional frontiers (Nataraja Guru 2001: 145).

References

Muni Narayana Prasad, Guru, 2003, *The Philosophy of Narayana Guru*, New Delhi: D.K. Printworld.

———, 1998, *Muṇḍaka Upaniṣad*, New Delhi: D.K. Printworld.

Nataraja Guru, 1987, *Vedanta Revalued and Restated*, Varkala: Narayana Gurukula.

———, 2001, *An Integrated Science of the Absolute*, 2 vols, 2nd rev. edn. (3rd rev. edn, 2020) New Delhi: D.K. Printworld.

Yati Guru, Nitya Chaitanya, 1987, *The Psychology of Darśana Mālā*, Varkala: Narayana Gurukula.

———, 1993, *Bhagavad Gita, A Sublime Hymn of Yoga Composed by the Ancient Seer Vyāsa*, 2nd edn, New Delhi: D.K. Printworld (1st edn published in 1981 by Vikas Publishing House, New Delhi; 3rd revised edn published in 2022 by D.K. Printworld, New Delhi).

Omana, S., 2022, *Vedānta, The Science of Consciousness: Sree Narayana Guru's Vision*, New Delhi: D.K. Printworld.

15

Element of Wonder in Guru's Philosophy

B. Sugeetha

NARAYANA Guru does not limit himself to the scope of Indian thought only, but encompasses all possible visions, anywhere and at any time. These visions are structurally strung together in his work *Darśana Mālā*, forming a symbolic garland enhancing the dignity of man through wisdom. He draws his inspiration from the Upaniṣads and the *Bhagavadgītā*, as is shown by his precise Advaitic terminology. Another important source of inspiration was derived from his own *tapas* (mystical discipline). Most of his works are the products of inner experience with a confirmation from outer textual sources. The present study is "The Element of Wonder in Guru's Philosophy" as elucidated in the second chapter of *The Philosophy of a Guru* by Nataraja Guru.

Alexander Pope said that the proper study of mankind is man. A person must have a complete philosophy which would help him to know his environment fully and globally, and also to live in it correctly by taking into account all his subjective traits and matching them with their counterparts in the objective world. Correlations, cosmic, psychic, or both have all to be matched through analogies or equations implying one or the other aspect, or both aspects at once, so as to reveal the full mechanics as well as dynamics of life. Guru's philosophy is meant to represent such an inclusive way. It is a healthy, wholesale and wholehearted philosophy in which dialectical counterparts are juxtaposed so as to yield a unitive vision of the Self in the light of the normative Absolute.

Indian Philosophy More Scientific

In order to mark the scientificity of Indian philosophy, Nataraja Guru put forward the views of Bertrand Russell and Max Müller. They hold different opinions about how philosophy is to become more scientific. Russell always rejected a wholesale approach to philosophy, and thus to Indian philosophy too. Max Müller, on the other hand, thinks that Indian philosophy is more philosophical than modern Western philosophies. He writes:

> What I admire in Indian philosophers is that they never try to deceive us as to their principles and the consequences of their theories. If they are idealists even to the verge of nihilism they say so, and if they hold that the objective world requires a real, though not necessarily a visible or tangible substratum they are never afraid to speak out. They are bona fide idealists or materialists, monists or dualists, theists or atheists because their reverence for truth is stronger than their reverence for anything else. – *Unitive Philosophy*, D.K. Printworld, p. 190

Georg Wilhelm Friedrich Hegel condemned Indian philosophy as adolescent and immature. Later, Max Müller, a scholar from his own country, Germany, examined both Eastern and Western philosophy impartially. His generous praise helps to counteract the aspersions cast on Indian thought by some Westerners. Müller's further remarks in the above context are therefore worth quoting:

> We know what an enormous amount of labour had to be spent and is still being spent in order to ascertain the exact views of Plato and Aristotle, nay, even of Kant and Hegel, on some of the most important questions of their systems of philosophy. There are even living philosophers whose words often leave us in doubt as to what they meant, whether they are materialists or idealists, monists or nihilists, theists or atheists. Hindu philosophers seldom leave us in doubt on such important points, and they certainly never shrink from the consequences of their theories.
>
> – *Unitive Philosophy*, p. 190

Indian thought endeavours to express the nature of necessary truth or absolute certainty. The Six Schools of Philosophy (Ṣaḍ Darśana) do not err on the side of the vagueness of speculation.

The definitions, though cryptic in linguistic formulation, are precise and clear-cut when properly understood. No doubt, Müller had in mind the six systems that have come down to us in succinct aphoristic formulations, when he applauds the merits of Indian philosophy as against Western philosophical speculation in general.

Scientific Status of Śaṅkara's Advaita

Advaita Vedānta attained a fully scientific status after Śaṅkara's critical restatement, following Bādarāyaṇa and Vyāsa. Here, the Absolute is treated wholesale and wholeheartedly in downright and outright terms of intellectual attack and emotional onslaught, to finally become fully apodictic. Emotions naturally resemble those of a mystic, while the intellectual approach corresponds to the pure vision of a man of dialectical wisdom.

This means that many methods of asserting different arguments can coexist in the vision of a *guru*. At the same time, they do not violate any law of thought. Modern Western thought, on the other hand, is a conflict between philosophy and spirituality. Those who are only familiar with Western thought will find the Guru's philosophical thought vague or just an expression of emotion. For them, intuitive vision or mystical vision would be too strong to gaze at compared to the twilight world of easy speculation in which modern philosophy has had to thrive.

In Guru's philosophy, speculation not only becomes red hot but attains to an incandescent white heat which reaches up to melting point at times. Nataraja Guru mentions in advance that those who think in a critical way should not consider this feature of the Guru's philosophical thought as a sign of the loss of critical thinking.

> Its value is only enhanced when it conforms to the requirements of a normalized absolutist method, a theory of knowledge and an axiology which does not exclude the overall human problem of minimizing suffering or bondage and of securing human happiness, individual or collective, from its scope.[1]

Such a vision of truth of any *guru* must be comprehensive, calm and flawless. It should not see life in a cold light. It does not fit the way of cutting problems piecemeal one by one and trying different solutions for each one. Nataraja Guru said:

---

[1] Nataraja Guru, "Philosophy of a Guru", in *Unitive Philosophy*, p. 192, D.K. Printworld, Delhi.

> Lukewarm, piecemeal, trial-and-error, ratiocinative, eristic, or sophistic approaches, dabbling in mere opinions and counter-opinions, do not serve the purpose of healthy, wholesale and correct speculation that Guru philosophy is meant to be.[2]

Thus, the bold speculative flights found in the verses from *One Hundred Verses of Self-Instruction*, referred here, may confuse the modern reader in respect of their normality, sobriety or criterial validity, from the standpoint of the ordinary reader.

Ascent Implied in Guru Philosophy

Nataraja Guru mentions some prerequisites for aspirants. Contemplation implies a form of ascent through intensity of thoughts. Passive and lukewarm attitudes are not conducive to yield the results of true contemplation. The inertia, the aches and pains of the body, love of easy-chair comfort are to be left behind by the aspirant. There is a factor called *tapas* (burning up) implying intensity of research, full alertness supported by clear memory building, willingness to give full attention to the subject or problem at hand and the power to listen.

An important prerequisite for one who wishes to grasp Guru's philosophy is hinted at in the very first verse of *Ātmopadeśa Śatakam*. It is a contemplative hymn or sacred scripture to be sung or chanted like the Vedas, Quran or Bible. It concerns the Self which is to be located neither inside nor outside the contemplator. Self-knowledge is to be sought introspectively with the outward going interests restrained and directed inward in terms of consciousness, which can be said to be both subject and object at once.

2 Nataraja Guru, "Philosophy of a Guru" in *Unitive Philosophy*, p. 88.

A mere academic interest or intellectual curiosity alone will not suffice for the task of self-realization. A wholehearted interest is needed. Ecstasy and wonder are only to be expected normally in the appraisal of such a high human value. An attitude of ceaseless adoration is recommended so as to attune the mind to the implicit central notion which is the content of the whole work. Such a notion, being beyond the paradox of logic, has to be approached dialectically. In such a unitive approach the attitude of reverence or adoration is but a natural corollary. An adoration of the Absolute as the highest and dearest of human values. The Absolute Self is an enigma and a wonder.

The Bhagavadgītā (II.29) states this unequivocally:

A certain person sees this as a wonder, likewise another
Speaks about this as a wonder. Another
Hears of it even as a wonder, but even
Hearing no one understands this at all.

How the visible world itself can be effaced in the full wonder of the Absolute in terms of meaning of sound is boldly touched upon in verse 52 of *Ātmopadeśa Śatakam*:

Filled with the word-content, that day the firmament shall radiant blaze,
And in it shall become extinct all the visionary magic;
Then too, that small voice completing tri-basic knowing
Shall cease and Self-radiance prevail.

One of the characteristics of Narayana Guru's work is that while being scientific, it is emotional also. In some places scientism is preferred and in some other places emotion gets highlighted. If we imagine that science is a vessel and the nectar filling it is emotion, we can see in some rare places this nectar overflows. This verse is one such example.

It is not possible to fully describe and clarify the meaning of this verse. Here Guru is reaching to the transcendental realm of *advaitānubhūti*. The insight contained in this verse will be revealed to each one depending on the extent to which one is ready to reach that level by holding the Guru's hand on the basis of his or her own meditation.

We should not leave out the two words *annāl* and *iṭam*. *Annāl* refers to time and *iṭam* refers to space. The visual world exists only in the realm of time and space. Time and space exist in knowledge or *aṟivu*. It is also an experience where time and space become inseparable in the unity of knowledge and Self-illumination.

The five elements, their qualities, time and space are all aspects of scientific thought. Without dismissing these scientific aspects altogether, they shine in the unity of knowledge and at the same time make that Self-illumination more scientific.

Guru writes verse 52 surcharged with inner experience hardly capable of being put into words. *Dhavani*, the word used by Guru, refers to the source of all visible realities.

The colourful world of attraction and repulsion in which we pass our everyday lives is brought under the grade of a visionary magic. When the higher level of perception or vision is established within consciousness by intense thought or contemplation, the lower vision tends to get weakened and abolished together, absorbed into white light full of meaningful import. Just as the vision of individual trees can get effaced when the forest becomes globally discernible. It is thus that the show of colourful magical display is said to be absorbed or extinguished in the higher through more interior vision. The vision gives place to meaningful sound culminating in the conceptual light of the absolute name. The still voice under reference here, which is the last link between outer and inner language, shall stop when the full vision of the Absolute is about to be established. This non-dual state is here compared to the All-Filling Light or Self-realization.

The element of mystery or wonder is pronounced is verses 35 and 65. Verse 35 is as follows:

Like the dawning all together of ten thousand suns
Wisdom's functions come: Such verily is what
Can tear asunder the knowledge-hiding impermanent
darkness of Māyā *here.*
And, as the primal Sun on high, prevails.

Here Guru makes more finalized statements about the way in which wisdom comes to the aspirant for Self-knowledge. Illumination as understood in the contemplative context is not one that takes place slowly and gradually. Within the relativistic context there might be accumulation of information about things or events which in time may mature to make the person concerned more and more worldly-wise, but the wisdom that has to do with the absolute asserts itself in quite another fashion. It is not a slow evolutionary process, but an overwhelming event in one's life.

Contemplative wisdom is different from ordinary accumulation of information about events, things or matters. It is something wholesale of an all-or-nothing character. This difference is due to the global totality of the subject matter of such wisdom which is none other than the Absolute. Wisdom that refers to the Absolute has to be itself of an absolute quality.

Verse 65 says:

There is nothing here that we have not already once known.
Hidden by form, Knowledge fails us; wakefully
To know all this there is none at all, limitless as it is;
Oh, who can there be at all to know this Wonder dear!

The reality that we attribute to the objects we see is to be traced to their sources by a process of reasoning which goes from effect to cause. Such a philosophical way of enquiry is natural and normal to the human mind. All things must have a cause and science is what reveals the cause behind effects which constitute all the appearances in which we live.

So there cannot be any effects without a corresponding cause. In the Indian philosophical context, we distinguish between *dharma,* the mode of expression of an object, and *dharmī* (the basic reality common to particular modes of expression which is the cause or agent that produces the effects or *dharmas*). This agent cannot be seen, but has to be inferred through the exercise of the faculty of reasoning. What hinders reasoning here is the visual aspect. Individual possibilities of wakeful knowledge are still more limited. Actual knowing, as distinct from inferential knowing, draws a still narrower circle around our range of vision of things. Even the outside wall of our living room is only known to us at second hand.

The "dear wonder" referred to in the last line is that aspect of the Absolute not subject to the influences of memory aspects of consciousness.

It has been pointed out, in verse 64, that memory is the enemy of spiritual progress. Only bold spirits can undertake the positive conquest of the unexplored aspects of what is known as *adṛṣṭa* (unseen) or *apūrva* (never-known before) aspects of the Absolute Truth, which alone gives crowning character to the notion of the Absolute itself. Here Guru touches upon a value that is Absolute and Supreme, and the most precious aspect of human wisdom itself. The last line declares how rare it is to attain such positive wisdom.

Here the limitlessness of the wonder is stressed to show that only the contemplative can have such visions.

16

The Path of the Gurus

Prabhavati Prasannakumar

The world today is wild with delirium of hatred
Conflicts are cruel and unceasing in anguish
Crooked are its paths.
And tangled its bond of greed.

RABINDRANATH Tagore penned these lines more than a century ago. The history of religion is one of bloodshed. The Crusades raged from 1096 to 1291 between Christians and Muslims to control the holy city of Jerusalem. In our own times, religious fundamentalism has taken the form of terrorism which threatens to destroy world peace. The only solution to save humanity from the grave situation prevailing in the world of today, caused by the artificial division of humankind, is to create awareness of the philosophy and teachings of Narayana Guru of the universality of human kind. He taught that mankind is of one race, one religion and one goal, which is of happiness. – One Caste, One Religion and One God for man.

The Narayana Gurukula is an important source to achieve the aim of universal peace and to ennoble humankind. Nataraja Guru was the most educated and enlightened person among the disciples and active followers of Narayana Guru. He applied the process of revaluation and restatement to purify Advaita Vedānta. The ancient teachings were simplified applying a scientific methodology making them relevant to modern society. Hence, Narayana Guru is the *ṛṣi* of the scientific age.

Nataraja Guru founded the Narayana Gurukula in the year 1923 with the support and blessings of Narayana Guru. The universal teachings of Guru were taught in living terms so that humanity can enjoy a peaceful and happy life. There are now branches in different parts of the world. Several Gurukula family members have obtained their PhD on the works of Narayana Guru.

May the ideal of universality of humankind and One World concept of Nataraja Guru flourish throughout the world! Narayana Guru's *satya darśana* (Absolute Truth) is called *aṟivu*. Our quest is to find "who am I?" and from where did I come? The goal of the search is to reach the Absolute Truth or the Self.

I am filled with gratitude and joy for the blessings that I received from the *gurus* of Narayana Gurukula to build up a meaningful and happy life based on *dharma*. My house itself became part of Narayana Gurukula. I had opportunities to visit different countries of the world to present papers on Narayana Guru's universality.

We can understand the Guru's teachings as follows:

The world is one by nature and divisions are only man-made.

1. One world – for us to live in.
2. Same air – to breath.
3. Water – to drink.
4. Food – to appease our hunger.
5. Moonlight – to remove the shade of night.
6. Language – for communication, and a Kiss, as reflection of love and affection.
7. The relation – mother and child.
8. Children – from generation to generation.
9. Sea – as the traditional heritage.
10. The sun – creating day and night.

When we look inward, we experience *ānanda.* We are in essence the Self or the Absolute which is *sat–cit–ānanda*. We have to become citizens of a world city where we can embrace humanity. This direct and simplified version of Vedānta reached the hearts of the common man through the earnest works of the great disciples and followers of Narayana Guru. The Narayana Gurukula's vision is to play such a role under the guidance of the *gurus*, who are great scholars and real *tyāgī*s.

As an inmate of the Narayana Gurukula and member of the Gurukula family, I feel it is my duty to emphasize the need for giving love and affection to children and directing them to lead a life of *dharma.* The success and achievements of my own life is the best example. In addition to the care given by the parents, children have to be led to *gurupāda*s (feet of the Guru), so that they can lead a life of *navākṣarī mantra* (*aruḷ uṭṭavanāu jīvi*): Grace, Kindness, Compassion or Mercy, Love, Sympathy. This is said to be the quality of man.

Being the most senior woman among the inmates of Narayana Gurukula, I am filled with gratitude for receiving the blessings of Guru. I now give a historical narration of my association with the *guru*s and Narayana Gurukula. As mentioned above, it was through my parents that I was brought to *gurupāda*. My mother K. Chellamma was a member of "Kumbolil Sabha" of "Varanappally Tharavad" where Narayana Guru had stayed. Hence I got a chance to play in the soil purified by the touch of *gurupāda*. My father was V. R. Madhavan, Vazhakkoottathil Kollam, an artist, musician and actor whose set the poems of Kumaran Asan to music. Narayana Guru blessed him when Guru took rest in Mangalam House, a branch of Shri S.M. Padmanabhan (Thottathil family), an ardent follower of Narayana Guru. He dug a public well in Uliyakkovil (Kollam) for the untouchables. It was inaugurated by Gandhiji on the invitation of Shri Padmanabhan, during Gandhiji's visit to Kerala and his meeting with Narayana Guru in Sivagiri. During my school days my father took me to Sivagiri every weekend. We packed food with us and travelled by the early morning train and returned by the afternoon train. After prayers, we sat under a tree and my father told us stories about Guru and sang poems of Asan, which I can recite even now.

I had an opportunity to listen to Guru's *prabhāśaṇam*" (speech), during the Jubilee Celebration organized in Sivagiri Sree Narayana College. By God's grace, I was taken to Narayana Gurukula by my brother Dr Sarasijan, who worked in the Gurukula Medical Mission, started by a Singapore-returned devotee and follower of Nataraja Guru. Since then I wanted to be in the Gurukula and finally Guru's decision favoured me to be a Gurukula inmate. The words and the love and affection of the *guru*s cleared the path to flourish in life and gave me mental power to be one with my kith and kin, neighbours and the society at large, to solve problems and maintain peace and happiness to the maximum extent that I could.

Nataraja Guru was the first disciple to take up the cause of Vedānta in the beginning of the history of Narayana Gurukula. Nataraja Guru wrote a Constitution for Narayana Gurukula.

What is known as Advaita Vedānta in India is revalued and restated by Narayana Guru. It is considered as an Upaniṣadic secret. Since it is thought that the mind cannot reach this secret, the opinion of great *paṇḍit*s was that we need not think about it. But this was not accepted by Nataraja Guru and Guru Nitya.

The secret can be explained with the help of a *mahāvākya*: *tat tvam asi* (*tat* = that; *tvam* = you; *asi* = That is only you).

This can be explained through a mathematical equation $(a + b)^2 = a^2 + 2ab + b^2$.

Though we understand that both are equal, what is understood as "sameness" is not written. Guru Muni Narayana Prasad explains, "All equations have two limbs with an 'equalness' sign between them. What we grasp is not the value or meaning of either of the two sides, but their oneness which is not written down. It is not possible to grasp this oneness without making use of these two limbs. In short, the meaning we grasp from what is written down is something that is not written down."

We understand *tat* = *tvam* (*satyam*), truth having universal dimension, by explaining two opposite sides. This method is called Dialectics. Truth is something which is neither universal nor individual and is to be found with the unseen eye. What cannot be seen with our ordinary sight is to be understood with our inner eye.

Nataraja Guru must be given credit for his insight into the dialectical secret hidden in the Upaniṣads. He was not presenting any new theory, but stated the method hidden in the Upaniṣads and used by the *ṛṣis*. Dialectical methodology has existed from the time of Plato in the West and was termed as the "coping stone of wisdom". The subject–object distance gradually reduces, and when I study about myself, subject and object become one. We are not able to go or think beyond that. That is pictured as "coping stone".

To study about myself, we must understand what is *ātmā*. By discussing both sides, one is able to intuitively see the *satya* (truth) which cannot be expressed in words. This is made known by explaining both the faces of it. By explaining two opposite sides, the secret is intuitively known without mentioning this truth. Thus, the *ātmā* of Advaita could be clearly understood with the help of the *mahāvākya*.

Natarajan was the son of Dr Palpu. Narayana Guru often visited the house of Dr Palpu. Guru used to have subjective discussions with young Natarajan. When Nataraja Guru took his Masters in Zoology and read a paper on Charles Darwin's theory of evolution at a science congress in Calcutta and returned to Guru, Narayana Guru asked Tampi (Nataraja Guru): "Is evolution still going on"? Tampi took it for a joke. Then Guru further asked,"If evolution is going on, why do we not see anywhere a monkey's tail dropping off?" On another occasion, Narayana Guru drew Tampi's attention to the reality of stones thrown by a Kuṭṭi-cāttan. Experimental knowledge shows that such stones were not available in the surrounding area. This phenomenon aroused Natarajan's interest to study it scientifically.

During his college days in Madras, Natarajan used to collect money and performed *daridra-nārāyaṇa-pūjā* (giving charity to the poor). Once, a day before he was planning an *annadāna* (giving food to the needy) Narayana Guru paid a visit and inquired about the preparations going on. Guru said that considering others as poorer than us is because of our mental poverty. This taught a great lesson to young Natarajan and thereafter giving food to the people was called *guruprasādam,* instead of *daridra-nārāyaṇa-pūja.*

Guru Nitya Chaitanya Yati and Mangalananda Swami were two great disciples of Nataraja Guru and with their presence Narayana Gurukula flourished and spread to different parts of the world. Mangalananda Swami's original name was Madhavan. His father was Haripad Pothappally Thevalasseryil Kunjusankara Panikkar and mother Arattupuzha Mangalam Edakazhiyil Velutha. He had one sister Mangalam Thekkeparambil Bharathi Amma and two brothers Shri Damodaran and Dr Vasudevan. His father was an artist and an expert astrologer, and he died before Madhavan was a year old. He was the youngest, and loved and cared by all in the family. After completing High School education in Karuvatta School, he went to Madurai for College Education. In the Intermediate Exam Madhavan failed which upset the elder brother Dr Vasudevan who beat him in anger. Madhavan felt sad and finally left the home without even informing his mother. He travelled throughout India, visiting almost all holy places and spent a few years with *saṁnyāsin*s in the Himalayan region. He was not satisfied with the knowledge he got from the *saṁnyāsin*s and thought of Narayana Guru.

Madhavan asked Narayana Guru for *saṁnyāsa dīkṣā,* but Guru asked him to go back home, finish his *karma,* leaving all relationship to receive *saṁnyāsa.* Thus after many years Madhavan reached Mangalam House. The ailing mother was looked after by sister Bharathi and brother in-law Shri Achuthan. The aged mother was longing to see Madhavan. She was satisfied and with a happy and contended mind she breathed her last. After finishing the last rites, Madhavan came back to Narayana Guru and received *saṁnyāsa dīkṣā.* As directed by Narayana Guru, Madhavan, now Mangalananda joined Nataraja Guru and along with Guru Nitya he nourished the Gurukula.

From the beginning there were many remarkable contributions from Madhavan, from writing, publishing, to reciting the poems of Kumaran Asan. He presented *Duravastha* and *Caṇḍāla Bhikṣukī* in the form of *Kathā Prasaṅga* which attracted the people. The *dakṣiṇā* he got from his presentations was used to feed the inmates of the Gurukula. Mangalananda Swami's final merger with the Absolute happened as he desired. In 1997 while speaking in a place called Piravantoor, near Pathanapuram, he fell down on the stage and breathed his last.

In 1956, when he got the family share, in Trichur, 5 (five) acres of land was given to Gurukula in Engandiyur (Trissur). Nataraja Guru, John Spiers, Mangalananda Swami together went to Singapore in 1956. A Gurukulam was started in Singapore, which exists even today. In 1970 *gurukula*s were founded in Palakkadu, Wayanad, Vaithiri and Tholpetty. When Guru Nitya was in America, his disciples founded two *gurukula*s. In Ezhimala Gurukula Nataraja Guru organized an eleven-day World Conference. At that time I was residing in a house Nataraja Mandiram, since no woman was allowed to stay inside the Gurukula compound in those days. A school named Gurukula One World School started functioning in Varkala Narayana Gurukula and I was a teacher there. One day when I visited Gurukula, Guru Nitya asked me if I could do honorary work in Gurukula One World School, I happily agreed. It was a proud moment when during the Oṇam season I received the so called "first salary" from the hand of Guru Nitya – one hundred and one rupee notes clipped together. Even now, after receiving UGC salary and other benefits, at the age of 77 the memory and how I had utilized it give immense satisfaction and happiness which cannot be expressed in words.

In the early years Nataraja Guru, his friends and their children, around ten, met in Narayana Giri (now *samādhi* of Nataraja Guru) in the evening for discussions on different topics. After food when they went to wash their plates Guru Nitya used to be with them, giving instructions on how to clean their hands and plates. They could also enjoy jokes from Nataraja Guru.

The present Guru, Guru Muni Narayana Prasad started staying in Gurukula permanently in 1960, taking care of various responsibilities. In 1960, both the author and Prasad Swami were there in the Gurukula. In the Convention of 1969, I was made the *adhyakṣa* (Presiding Officer) of the "Gurukula Balalokam" meeting. Guru Nitya introduced me to the audience saying, "You may think, why this big girl is made the leader of this Balalokam. I did it because I know that her mind is that of a child." Among the audience were people from Konni and they requested Guru to include me in the first convention in Konni Ashram. I was fortunate to be with Guru.

Once when Guru Nitya was director of Psychic Research Institute in Delhi, I got a letter from him informing me of the arrival of Nataraja Guru, asking me to go and meet him. "Guru's nature is such that he may get angry sometimes." It was a happy moment when I met Nataraja Guru in Narayana Giri. At that time there was only a hut at the Giri. Nataraja Guru asked me about my postgraduate subject. On hearing Political Science he asked me if I had heard of Rousseau's works. He was pleased when I mentioned *Social Contract* and *Emile*. I was present every morning for Guru's classes as a disciplined student.

One day during the 1969 Convention, Guru Nitya called me and asked me to go to Trivandrum Medical College Hospital since my mother was admitted there. After two days Guru came by car and took me from the hospital. A lady was sent by Dr Sumangala to take care. I was taken to Varkala Gurukula. After spending nearly an hour, Guru sent me back after giving a newly-printed book to read. After one or two days Nataraja Guru came to the hospital with three of his followers. Before going back Guru said to my mother,"In two days you are going to be discharged and go home. In the near future you need not be in the hospital. If necessary after years you can be in hospital for two days" and gave her an orange. It is a wonder that Guru said this in the year 1969, and mother died in 1995. Nearing her last day she was in the hospital for two days (after twenty-six years).

I am what I am today as the result of the love, affection and care given to me by my *gurus* and my father, who guided me to *gurupāda*. My philosophical foundation was laid by Nataraja Guru and it flourished under Guru Nitya and continues under the guidance of Guru Muni Narayana Prasad. If we take the 100-years history of Narayana Gurukula, the period of Guru Nitya can be considered as the Golden Age of Narayana Gurukula. He lived the principles of the *navākṣarī mantra*

– mercy, kindness, sympathy, compassion. Anybody could approach him with his/her problems. I know of one case in which a colleague's daughter, facing the problem of dowry, was contemplating suicide. I asked her to write to Guru Nitya. Within a week she got the reply and the first sentence was enough to give her peace of mind: "Is a girl to be auctioned like a cow in a market"? Nowadays we hear of suicides and cruel murders in the name of love affairs. Such tragedies happen in the absence of *guru*s who give good advice and take the initiative for solving problems.

Narayana Guru gave humanity the Principle of Universality "One caste, One religion and One God" for mankind. Nataraja Guru along with Garry Davis gave us the notion of "One World" and "One World Government" for maintaining World Peace. Nataraja Guru met Garry Davis on a ship while crossing the Atlantic Ocean. Garry was an ex-American pilot who bombed Germany during the Second World War. When he realized the suffering he had caused, he had a complete change of mind and renounced his US citizenship and declared himself a world citizen. Nataraja Guru fully supported Garry Davis and even signed his World passport. A symbolic World Government was declared on 4 September 1953. Nataraja Guru also developed the theory of "One World Economics". Guru Nitya was acting as the Commissioner for "One World Education".

Nataraja Guru and Guru Nitya took the necessary steps to groom Guru Muni Narayana Prasad to be an efficient successor. Between Guru Nitya and Swami Narayana Prasad all the major Upaniṣads commentaries were written. With *Gurukulam* monthly, the other publications also increased a lot. Nataraja Guru organized a meeting in 1971 which was to be his last. The author got an opportunity to attend the meeting as the representative of Gurukula. Nataraja Guru suffered a stroke in 1972 and was bedridden due to paralysis in a hospital in Bangalore. Immediately Guru was brought to Varkala and attained *samādhi* on 19 March 1973. Guru Nitya immediately returned to Varkala. He collected all the keys and gave them to Guru Muni saying, "Hereafter this is your responsibility". The responsibility is being carried out perfectly.

Guru Muni Narayana Prasad's original name was M.N. Prasad. Nataraja Guru said let there be no change of initial and called him "Muni Narayana Prasad". He was born in the year 1938 in a place called Thanneerkonam, Alathukavu in Nagaroor. His father was Madhavan, Headmaster in a Primary School and mother's name was Narayani. He lost his mother in his early days. From when he was 7 years of age his father took him to Sivagiri and other enlightened gatherings. After completing his school education, he joined the Civil Engineering Certificate course in M.V.T. Institute of Civil Engineering in Maruthamvilakam, Vakkom and passed with first class. He worked both in Kollam and Trivandrum. After the passing of Mangalananda Swami he had more responsibilities in the Gurukula. Nataraja Guru asked Prasad Swami to resign his job and become fully involved in Gurukula affairs.

Narayana Guru had envisioned a Brahmavidya Mandir. But by 1973 only levelling of the ground was done. It was completed later by Guru Nitya at the *samādhi* place of Nataraja Guru. It is named The "East–West Universe of Brahmavidya". The beautiful spacious building has space to conduct meeting on the first floor and a library on the ground floor.

Nitya Chaitanya Yati's name was Jayachandran. He was born on 2 November 1924 in the place called Vakayar near Konni in Pathanamthitta district. His father was Raghavapanikker, a government school teacher and poet and mother Vamakshi Amma, daughter of Thazhatheril Kesava Panikker. He had a brother – Balachandran – and two sisters (Subhashini and Dr Sumangala). Before her marriage Vamakshy Amma happened to visit Narayana Guru and got blessed. The radiant light witnessed on the face of Narayana Guru deeply impressed the mind of Vamakshy Amma. The 18-year old girl started regular prayers promising the Absolute, "If I am blessed with a son, I will surely make him a *saṁnyāsī,* like Narayana Guru". On 2 November 1924 the child was born and from early childhood Jayachandran showed the spirit of a *saṁnyāsī.* At home the family used to read *Buddhacarita* in the evening. When he was 10-year old, Jayachandran asked his mother whether he could be a Buddha and mother responded positively. Instead of being a mother, who never wanted her son to be a *saṁnyāsī,* his mother was very happy and supported him in all possible ways to develop the way of life of a *saṁnyāsī.*

For some time he led a wandering life and had to face dire situations. Once, hungry for food, he inadvertently enrolled in a group of Air Force trainees. His chance meeting with John Achan (teacher in Union Christian College) was a turning point in his life. Father John helped him to enrol in U.C. College, Aluva. He then obtained his Master's degree in philosophy from Trivandrum University College. He took *saṁnyāsa* in the presence of Ramana Maharshi at Tiruvannamalai. After meeting Swami Mangalananda they became close friends and the disciples of Nataraja Guru.

Guru Nitya could easily handle any subject under the sun, and even a number of languages. He was also a very good artist, musician, poet, photographer and cook. Once I asked the V.C. of Open University: "If you are asked to say five sentence about Guru Nitya, What will you say?" The answer was, "He was beyond everything: If Guru – he was beyond a Guru, a friend – beyond a friend."

After completing chapterization of my PhD thesis, I decided to show it to Guru, even though my guide had given a very good opinion. Guru said, "This is incomplete. For a good thesis you have to add one more chapter – The Origin of All Political Parties of Kerala". The title was "Inter Party Relations in the Coalition Government in Kerala in the 1980s". My thesis happened to be the first to include the origin of political parties under the University of Kerala.

Guru enjoyed each and every aspect of nature, taking personal interest in keeping a beautiful garden too. Due to a stroke, Guru had to take some Ayurvedic treatment and for that he stayed at the Kanakamala Gurukula for some time. From there Guru went to Ooty Gurukula and began to follow his daily routine.

The author talked to Guru in the night of 13 May 1999 at about 8.30 p.m. Guru Nitya asked, "Mone, when will you come?" "You must be here at the earliest." I found myself in Ooty the next day.

Fathima Beevi, the first lady Justice from the Muslim community, was appointed as the Governor of Tamil Nadu. She wanted to meet Guru Nitya and an appointment was fixed for 14 May evening. Once Guru said: Nowadays it is difficult to get government sanction to have one's burial in the personal spot. If there any difficulty in getting permission to have *samādhi* in the Gurukula compound, immediately take the body to the cremation ground and take the ashes.

The presence of the Governor in the Gurukula made it easy to get permission to have Guru's *samādhi* in the Ooty Gurukula compound itself and a *dhyānamaṇḍapam* was built by Thampan Swami (now Swami Tanmaya). Thus comes the responsibility of keeping all the keys of Gurukula in the proper places, to the present Guru Muni Narayana Prasad. Since he was made strong enough and properly trained by the two great *gurus*, he could properly guide Gurukula by organizing study classes, running a boarding for students, weekend classes in Gurukula after prayer, publication of *Gurukulam* magazine properly, etc. We are now celebrating a year-long centenary of Narayana Gurukula. This piece of writings is a part of centenary souvenir. At present Thyageeswara Swami and Tanmaya Swami are ready to take up any work to help Guru Muni.

Anyone interested in acquiring knowledge (*aṟivu*) is welcome to stay and study in the Gurukula, get satisfied and be happy. At the Gurukula one acquirers knowledge on the basis of the simplified Vedānta of Narayana Guru. The motto is One caste, One religion and One God for man. We may say that Gurukula nurtures the wisdom of the Upaniṣadic *r̥ṣis*. Unitive understanding includes all the different subjects, or sciences, and the essence of all religions. Those who love a unified approach to life and learning can benefit from the Gurukula.

17

Guru's Discourse in a Post-Liminal State

Shareena Banu

ON THE vast canvas of ancient Indian wisdom, nature sketched the portrait of a God-like Guru. He embodied the essence of the timeless wisdom of all his predecessors and successors. Comparatively, the awe-inspiring life stories of the *gurus* bore striking resemblances to one another. The narrative presented on a *guru* in the current piece aligns with the familiar trajectory and purpose of their life.

An ascetic lived in the wilderness surrounded by untamed wild animals. He meditated beneath the canopy of a towering banyan tree, in search of truth. Beside him, the sublime eternal river meandered, washing away all his intense emotions. Nature, a master sculptor, worked through him, refined him like those lustrous pebbles scattered all over the riverbank, and bestowed on him serenity and equanimity. For two to three years, he nestled in the same verdant ambiance, immersed in contemplation and absorbed in eternal silence. At the heart of that mystique valley, he dwelled like a living, picturesque poem. The beautiful serene valley soon absorbed him, making him part of its myriad hues and shades. And then the inevitable happened: he went through a metamorphosis, and nature, ever welcoming, integrated him as a delicate flower petal of that eternally blossoming "time-lily".

As time passed like a calm stream, one day deep down from the recesses of his silence, he heard a melodious bird singing the hymn of *Āraṇyaka*. Fluttering its wings, the bird emerged from within him and took flight into the vast expanse, and then briefly perched on the banyan tree and then soared across the sky. Gently descending, it dipped its wings in the holy river, before continuing its journey towards the horizon.

At this very instant, his identity dissolved and seemed to be converged at a singular dot. At this point, his gaze touched and synchronized with the silvery sheet of water flowing, the bird ascending to heights and the vibgyor sunbeams. In that threshold moment, in front of him, time extended back and forth, time sculpting itself, enabling him to transcend all dividing lines or boundaries, and accomplishing an absolutist notion of time. He suddenly experienced a surge of energy, coursed through him, running in him that profound and pristine vital energy, enabling him to transcend the threshold of the liminal or "in-between" state, and achieving post-liminal or "non-dual" luminosity. And he thus surpassed the hard and unbearable liminal stage of humans. The liminal stage could be compared to twilight, swaying between day and night. Or it is equal to the push–pull experience of an unborn baby while waiting inside the mother's womb, before making a grand entry into the world. Or someone was disturbed and sad for long and relaying too much on the mind to find solutions, leading to overthinking and hyperactivity, resulting in losing their natural intuition. They are either overconfident or utterly uncertain of their actions. In this liminal space, one was unable to reach the "the correct intuitive way of thinking", however, it occurred intuitively to a transcendent *guru* who had already traversed this limbo terminal status. (Having said that, there were multiple truths and hence multiple correct intuitive or absolute ways of thinking. Apparently, it explained to us why different *gurus* spoke differently of truth; Pure Truth was seen simultaneously as One and Multiple.) Beyond that liminal level, the state of the enlightened was identical to a bud that had blossomed. In that post-liminal state, he touched new heights of creativity and Oneness. Self-actualization soon came to him. What is more, mother earth showered celestial flowers on him as blessings.

The first part of the life of the Guru thus ended with his enlightenment, and we ventured into the next part of his extraordinary life.

Seated on a mountain peak, as fully awakened, he was surrounded by a small number of genuine wisdom seekers who gathered there travelling from far afield. They were his devoted disciples. The presence of the dedicated disciples set the atmosphere for the transmission of knowledge and also for he knew his life's calling (*svadharma*) to be a *guru*. His responsibility or rather destiny was to enlighten the world steeped in utter darkness. Intending to dispel the darkness, he started a *gurukula* for the education of his *śiṣyas*. Inspired discourses emanated from the Guru, in a state of post-liminality, in order to equip the keenly attending wisdom seekers with spiritual maturity.

In one such discourse, the Guru eloquently rendered a discourse on the bipolar relationship between the *guru* and the disciple, a central theme of his Advaita philosophy of non-duality. The enlightened master forged an appealing comparison between the sublime stature of the mother and a real *guru*. In order to be able to better fathom it by his devotees, he vividly expounded it, entirely, by way of illustrating the disciple's complex inner nature and inner journey, comparing it to that of the stages of a small baby growing up.

He described how soon after the birth, the baby started crying, yearning for the attention of the entire world and particularly of the mother, often unconsciously. In the same way, a disciple, who was beset by suffering in his life and was consumed by total despair, and had stopped believing in worldly pleasures and the extraneous realm, experienced long unhappiness and longed, unconsciously or silently, for a spiritual *guru*, who deceptively lurked somewhere or everywhere.

Though difficult for others to discern the true reason behind the clamour of a newborn baby/true seeker, it sounded rather spontaneous and resonated with the mother's heart and was recognized easily by the real *guru*. Only a true *guru* fathomed the true meaning of human suffering. The overt and covert spiritual clamours were the signs of spiritual turbulence in a person, longing intensely for nature to open, for him, its creative window. Though others often were clueless about the source of the cause of human sorrow and unhappiness, and sometimes also misjudged the true wisdom seeker's clamour, in reality, it was the vociferous outpouring and divine vocalization of his existential crises, making his liminal presence felt to him and others. It poured out from the realm of the unknown instinct, natural to all humans; but not all would term it as spirituality. It was indeed a rare moment, since he enthusiastically attempted to affirm his beingness by inviting us all to his inner world, which was later unraveled only by a real *guru*.

As the dictum goes, the *guru* appeared when the disciple was ready to commence his spiritual journey. This unique idea was widely spread across the mystical world and true in this case. The *guru*, soon after finding his true disciple, like the mother who took care of the child's well-being, took responsibility for the spiritual well-being of his disciple. A complex scenario encompassing a range of emotions followed, the *guru* alternated between displaying extremes of emotions; for instance, sometimes bestowing care, concern and compassion on the disciple and at other times expressed hardheartedness and even anger.

Once the infant grew older and reached the age of three or beyond, it tended to exhibit temperamental outburst more often, especially, in one particular instance. It happened when the mother, all dressed up, was prepared to go out. The child immediately and instinctively sensed a crisis and became anxiety stricken, anticipating the separation from the mother. The thought arose: she is going out, leaving me alone here; perhaps she will abandon me forever and I will be all alone. As soon as she left the room and stepped outside, the child followed, crying. It was that emotional outpouring, the moment of a strong feeling of solitariness and poignant expression of human attachment. As the mother further moved away from the vicinity, the child almost fainted. Similarly, the *guru* might also leave the disciple or even asked him to leave, in either case, it would be painful for an emotionally dependent or attached disciple. We could only imagine here his overwhelming sense of disappointment and the subsequent events that could have followed, constituting a dramatic plot; which could not be described here due to paucity of space. The mother and child are tethered to each other by a divine bond. So are the *guru* and the disciple especially in the initial stages of their relationship. The disciple remained psychologically tied to the *guru*, and often felt insecurity and precariousness.

For instance, the physical presence of the mother/*guru* had a palpable effect on the child/disciple. He imagined the bearing of her/him parting with him. The very thought of being deprived of the presence of the mother to the young child or the *guru*'s disappearance to the budding spiritual aspirant, who was emotionally vulnerable, was akin to a hypothetical situation where one was throttled and deprived of breath. It was comparable to a bird, suddenly caught and caged by a cruel hunter, grieving over its denial of freedom. It was similar to a poet who suddenly lost his vision and missing access to the source of all the enthralling moments of his kaleidoscopic world, robbed of the visual marvels he once cherished, such as the light rays radiating the morning dew, a sunset on an exotic beach or a glistening ruby.

In the early stage, the infant/disciple experienced a hierarchical vision. Initially, the infant refused to substitute the mother with any toys or other enticing objects. For him, the mother/*guru*, seemingly, was the only appealing and beautiful thing on earth. Gradually, as the child grew older, the mother's constant appearances and disappearances provided the child with the unconscious reassurance that she was always there. Hence, the relationship ripened and became enduring, and the pangs of separation diminished. The same applied to the reciprocal relationship between the *guru* and the disciple. The *guru*'s presence pervaded. He gradually grasped the necessary dialectical structure of his *guru*'s thinking. The *guru* nurtured him in order to transform into a new independent spiritual being. The *guru* imparted the science of the Absolute (*brahma-vidyā*) and prepared the disciple to receive the wisdom on Ultimate Truth. Under the guidance of the *guru*, he eventually became a young *yogī*.

Now fully matured, the disciple became freed from all attachments, and the earlier tormenting dialectical movement between attachment and detachment found in his *guru*'s behaviour became normalized and understandable. In other words, at this juncture, he learned to be neither attached nor detached; in Advaita teachings, it was called realizing *indriya vairāgya,* wherein the attachment of the *indriya* to all the sense pleasures ended, and total detachment was achieved, leading to the cessation of perpetual suffering. Having reached this necessary tipping point, the disciple became a supraliminal optimist, consciously and incessantly reassured of the invisible presence of the vital life-force within and without. As this belief became deeply engraved and embodied within him, he regarded it as a natural inner propensity and disposition of himself. The *guru* was eventually supplanted by unwavering certitude in the Absolute and gradually touched his *ātman,* attaining Self-realization and enlightenment.

Epitomizing the timeless *guru*s, thus ended the *guru,* his profound and inspirational discourse on attachment and detachment, and the disciples wholeheartedly received it. Embracing this eternal wisdom, they practised it to the utmost and strived to live up to their *guru*'s teachings. One disciple was particularly inspired by this educational philosophy of his *guru* and established a new *gurukula* based on the teachings of his *guru,* in a verdant hillock. And thus the tradition of *guru*s and their timeless wisdom lessons continued to this date.

Thus, the story reached its conclusion, with a new *guru* succeeding the other. Sometimes a true *guru* appeared centuries after to address human sufferings; nevertheless, the wisdom imparted remained the same; however, revalued and restated.

18

A Glimpse of Indian Spiritual Evolution

Swami Mangalmurti

THROUGH the immense course of time India has witnessed various stages of spiritual evolution beginning from the Vedic age. The *ṛṣi*s were spellbound by the paramount powers of natural phenomenon, such as the sun, thunder, fire, air and water. These seemingly ordinary events in day-to-day life, inspired the *ṛṣi*s with imagination and amazement. They surrendered themselves to these powers and began to pray to them regularly. There was a sort of awe and fear in the minds of the *ṛṣi*s as they adored these nature-gods. We can trace the origin of *yajña* to the offerings made to please the nature-gods. For Vedic *ṛṣi*s the powers of nature were godly manifestations. Western Indologists have named this polytheism.

With their keen observation of nature the *ṛṣi*s reached the conclusion that there is only one paramount superpower, which is invisible and governs all natural phenomena. From this the concept of monism developed later, *eke devaḥ sarva bhūteṣu gūḍhaḥ* in the Upaniṣadic text.

The two streams of polytheism and monotheism have run a parallel course in the Indian spiritual atmosphere. The practice of making offerings through *yajña*s (fire sacrifices), to the nature-gods with the assumption of one paramount power hidden in all natural phenomena became prevalent. Emphasis of *yajña* was on *karma*, whereas monism adopted *upāsanā* and *jñāna* as the main means for spiritual growth. Consequently in India we see two types of lifestyles – worldly life and life of renunciation. But India's worldly life has never been materialistic living in the modern sense. The Indian people adapted a four-step path towards holistic living, which is *dharma, artha, kāma* and *mokṣa*. It seems quite unusual that some royal figures of Vedic age, like King Janaka of Mithilā, were monistic in discipline, aspiring for the knowledge of Ultimate Reality along with leading a detached life. This has been a unique feature of Indian spirituality.

On the other hand, the desire of kings to confirm their places in the heaven and the greed of priests led to elaborate *yajña*s which became expensive and out of reach of subaltern people who could not afford to perform these *yajña*s. Ignored and forgotten these people turned towards Buddhism.

Emergence of Buddhism was a landmark event against the authority of Vedas, followed by opposition to the *yajña*s. The Buddha succeeded in his mission and a new religion was founded. Desire (*tṛṣṇā*) or craving was identified to be the main cause of human sufferings. The ideology of the Upaniṣadic monastic life of renunciation and the Buddha's idea about life stand parallel to each other.

Though Gautama Buddha proclaimed the middle path as his ideal mission of preaching, his Dhamma was basically renunciation-oriented. Though Buddhist monks and householders fulfilled each others needs, the division of society in two groups was clear. Mahāyāna sect emerged to bridge the gap between the two groups. Buddhism reached its zenith during the reign of Aśoka, the emperor. The fifth century is treated as the Gupta age in the history of India. During this period, some aspects of Buddhistic and Aryan traditions intermingled with each other. Still, Buddhist philosophy was profound and there were great Buddhist philosophers living in those times. In the course of time, Buddhism began to gradually decline. During the eighth century, Ādi Śaṅkara was able to challenge Buddhist doctrine with his argument of the witness (I-ness), the omniscient, omnipotent and omnipresent *ātman*.

Śrī Rāmānuja and other *ācārya*s elaborated the monistic idea of *ātman* (*Brahman*), the Ultimate Reality, to be constituted of *Īśvara*, *jīva* and *jagat*. The concept of *māyā,* taught by Ādi Śaṅkara, became very popular and was used as a device to reveal the meaninglessness of worldly life. The concept of *māyā* was accepted by all later *ācārya*s, and the word *māyā* became so popular that it has become a cliche of Indian folk. *Māyā* affirms the falsity of worldly life and clears the path for a genuine seeker towards the goal of realization.

In the long span of time Aryans and Buddhists came close to each other. Buddhist monasteries were well organized and regulated. Aryans were not like that, a difference that can be seen even today. It is the plus point of Aryans that they have opened the sky to get fresh air of spirituality. As far as monistic philosophy is concerned, the great seer Narayana Guru is a pure Absolutist. In the practical field of life he accepted the challenge of social reforms without which his mission could not be accomplished. He built temples where important Hindu deities were installed and worshipped. He established many schools to enable a new generation to come out from the abyss of ignorance. Phenomenal (perceptual) and noumenal (conceptual) worlds stand side by side. We cannot separate them. A proper balance between the two is necessary for a holistic life and for the pursuit of our ultimate goal, that of liberation from the bondage of earthly desires.

19

Narayana Gurukula
An Invaluable Bliss

Deepanjali

How can one describe what love is? How can one describe what beauty is? How can one describe what ultimate value is? Think of an empty stage that reverberates the undeniable presence of an invisible actor. That is what happens to your life's stage when you liaise with the Narayana Gurukula. It does not exert itself in your life, but you realize that it has permeated every aspect of your life. There is no coercion, but you get spontaneously drawn into the nectar of *ātma-vidyā* that is found at the very centre of the Gurukula. Yes, you are drawn to it like a hungry bumble bee, who realizes that this flower contains the nectar of all nectars.

It was on an evening of the early 1980s (I do not know the exact year, could have been 1983 or 1984) that my enthusiastic mother took me and my elder sister to see an important person who was visiting Edapally in Ernakulam. The person was none other than my dearest Guru Nitya Chaitanya Yati and the venue was none other than Narayana Gurukula at Edapally. Even as a toddler I could sense that this place was different from all other places I had been to. The atmosphere of the Gurukula, just like any other Narayana Gurukula, was suffused with nothing but a deep silence. The silence was periodically broken by words of love and words of wisdom import. This was not a place for the usual small talk. Although there would be many faces you would be seeing for the first time, none would qualify for being called a stranger. All were considered *aparan* (not other) person. I was surprised to find that this is a place where people from all walks of life, irrespective of gender, caste, race, religion, social class or any other superfluous compartments can come together and realize their oneness. No one will order you around and you shall not attempt to control others. It was an oasis of sheer oneness of all, amid the arid desert of "you and I".

No words can describe what it is to be in the presence of a wisdom teacher who considers every soul as his dearest darling. Even as a small child I used to have my share of burning family problems and associated anxiety. However, all my restlessness would simply vanish without trace whenever I saw Guru Nitya's lovely face and heard his tender voice. Guru was a living wonder, a fountainhead of knowledge, an unmatched expositor of word wisdom. People were drawn to him like ants to sugar crystals. Guru's presence ensured that the needs pertaining to both your *bhukti* and *mukti* were taken care of.

Later, I had the good fortune to visit Ooty Gurukula in Fernhill, Nilgiris. A heavenly place in the true sense of the word. Nested among the scenic Western Ghats mountains, the Ooty Gurukula is an abode of peace, wisdom and beauty. We used to attend morning prayer followed by Guru Nitya's class on any of the numerous subject matters of wisdom import. By afternoon a simple but tasty lunch awaits you in the Gurukula kitchen. I have always felt that there is something special about the food that you partake in any Narayana Gurukula. Being laced with Guruhood's unfaltering love, the unassuming rice *kāñjī* accompanied by one or two side dishes, mostly sourced from Gurukula's own kitchen garden, could put a sumptuous lunch from a luxury hotel to shame. I cannot forget how once Swami Tanmaya made a dish out of chow-chow leaves. It definitely fulfils something more than the hunger pangs of your stomach.

I have attended a few Gurukula conventions, both in Ooty as a child and at Varkala Gurukula as a grown up. Gurukula conventions are elaborate wisdom feasts. Formulated and implemented according to the schema prescribed by Gurukula's founder Nataraja Guru, these yearly festivals of knowledge are a rare opportunity for all attendees to delve deep into the various knowledge traditions that have originated in various parts of the world. Any system of thought which attempts to bring us nearer to the Ultimate Truth is given its due place in this scheme. Learners affiliated to the Gurukula curriculum on the Science of the Absolute present their research papers in these seminars. This is followed by further deliberation and discussion on the topic actively involving listeners and led by experienced moderators. Again, it is amazing how the modest space and accommodation available in the *gurukula*s never fall short for anyone who understand that there is more to life than seeking creature comforts. During the COVID-affected years, the Gurukula seminars were conducted as online webinars with participation by interested learners from various parts of country as well as abroad. This could happen due to the selfless efforts put in by a few diligent *brahmacārī*s of the Gurukula who realized the potential of the digital world to reach out to all those interested to explore the principles for which the Narayana Gurukula stands.

The *Gurukulam* magazine in Malayalam is yet another cardinal part of my Gurukula experience. I got "exposed" to it even as a child. At that time, I could not understand any content other than what was printed in the children's story section. Over time, however, as I gained grasp over the basic principles of Vedānta, the magazine articles began to make sense to me. The preface article (*mukhavura*) written mostly by Guru Muni Narayana Prasad would help one understand Narayana Gurukula's stand on the contemporary international, national or any other important subjects. Every issue of the magazine will also contain a poem by Swami Tyagi in which the *ānanda* essence of the Ultimate Truth (*ātman*) bubbles out as verses of sublime beauty. One has to read the life stories of all the successive *guru*s of the Narayana Gurukula *paramparā* to understand the hardship that they have endured to seamlessly publish this magazine which is modest in appearance but elegant in content. The magazine is an apt example to Gurukula's resilience against the vagaries of time.

The Narayana Gurukula literature is one of its kind. It can be unequivocally named the "treasure trove of wisdom". Firmly based on Advaita Vedāntic wisdom as envisioned by Narayana Guru, the works of all successive Guru's in the *paramparā* as well as that of other disciples are found here. Any ardent student of *brahma-vidyā*, who belongs to the present age of science, need not to go anywhere else searching for both basic and advanced learning content. In addition, there are books with content pertaining to humanities, economics, art, aesthetics, psychology, poetry and children's literature, to name a few. Elaborate commentaries on Narayana Guru's works as well as traditional Vedāntic wisdom texts are available, which were written keeping in perspective the modern mind.

I am one of those blessed people who have received invaluable letters from Guru Nitya which will invariably end with "love and blessings". I used to write short poems in English as a child and send it to Guru Nitya and he would painstakingly do the needed corrections. In one of his "reply poems" he had promised to "hug me tight forever". Indeed, in the form of Muni Guru and his successive lineage, I continue to enjoy the privilege of being protected by the Gurukula shield. I could write pages and pages about the numerous instances in my life where Muni Guru or Swami Tyagi would have given me unconditional emotional support. Above all that, the time spent in the Gurukula has taught me that there are no "absolute sufferings", ignorance is the cause for any suffering. Life is an ocean of *ānanda* and it is me who has mistaken it for an ocean of suffering (and happiness as well!). I have seen how the *guru*s and *swami*s in the Gurukula endure their grave and physical ailments with equanimity. The secret behind their equanimity is one and the same secret behind me too.

Everything about the Narayana Gurukula is equipoised on an invisible middle ground which was described by Narayana Guru as the *nadunila*. If one has lot to learn from Gurukula's actions, then much more is to be learnt from its inaction. A day spent in the Gurukula recharges one for great many days in one's daily life. The rejuvenation is triggered by one's self only, which finds itself "at home". For me, Gurukula is the sanctum sanctorum of my heart. The following verse of the *Gītā* (II.40) aptly describes what to expect if one associates with Gurukula with an open heart:

> *ne 'hā' bhikrama-nāśo 'sti pratyavāyo na vidyate* ।
> *svalpam apyasya dharmasya trāyate mahato bhayāt* ॥

In such (a path) there is no forfeiture of any merit nor is there involved any demerit by transgression. Even a little of such a way of life saves one from great apprehension.

20

Why Zen?

An Intro to How? Why? And What? about Zen Practice and How It Can Be Beneficial in an Indian Cultural Setting

Gita Gayatri

MAN has been in search for what is real in him all through the ages and yet it still remains a mystery. There have been philosophical thinkers throughout history who have made significant contributions to human thought and understanding. Socrates, Plato, Aristotle of ancient Greece, Thomas Aquinas of the medieval period, René Descartes who ushered the modern period with his famous statement, *cogito ergo sum*, "I think therefore I am", Emmanuel Kant, known for his *Critique of Pure Reason* which had a profound impact on metaphysics and epistemology, Friedrich Nietzsch of the ninteenth century, Jean-Paul Sartre and Simone de Beauvoir of the twentieth century, contemporary thinkers like Noam Chomsky, the fathers of modern linguistics, there is an endless list of thinkers in the world.

Immersed as we are in the fast-paced world, there is need to pause and reflect, "Who am I", "Where am I headed to?" and What is this world and how can I connect with the world in a way that gives meaning to my life". There is an ongoing search and thirst. What we believed would give satisfaction and contentment, seems to run out after a while and we seem to be searching again and again, despite all our efforts to find permanent solutions in this impermanent world. At some point in life everybody stops and starts questioning the meaning of their existence.

In India, spiritual practice has continued unhindered for thousands of years. Of course, the British ruled India for 200 years, and most of Indian art and culture was broken, the very backbone of Indian culture was broken, but in the realm of thought, India has always had freedom. On the contrary, Europe has a history of intolerance, and even torture, which peaked during the Roman and Spanish Inquisitions between the twelfth to nineteenth century. People were punished and tortured severely or prosecuted or exiled for non-conformity with the Catholic doctrine. There was no freedom of thought. Hypatia of Alexandria, a neoplatonist philosopher, astronomer and mathematician, was brutally murdered by a mob of Christians in March 415 CE. Giordano Bruno, Italian philosopher, astronomer, mathematician and occultist whose theories anticipated modern science, was burnt at the stake on 17 February 1600 in Rome. History is replete with such extreme suffering in many parts of the globe.

India, on the other hand, has an unbroken lineage of thinkers and philosophers from ancient times to the present day. The history of thinkers in India is rich and diverse, spanning thousands of years. India has a vast tradition of intellectual thought that encompasses a wide range of fields and is the root of many schools of thought. Ancient philosophers like Veda Vyāsa, Gautama Buddha, Ādi Śaṅkarācārya, medieval scholars like Ācārya Rāmānuja, mystics like Kabir and Guru Nānak, modern thinkers like Rabindranath Tagore, contemporary philosophers like J. Krishnamurthy, to name a few. Indian philosophy has a rich history of diverse schools of thought. Some of the prominent schools are Nyāya, Vaiśeṣika, Sāṁkhya, Yoga, Vedānta, Buddhism, Jainism, Sikhism and Śaivism. Some are atheistic and some theistic. These schools of thought have shaped the intellectual and spiritual landscape of India for thousands of years, and they continue to have a profound influence on Indian culture and philosophy.

Neo-Platonists believed that human perfection and happiness were attainable in this world, without awaiting an afterlife. Perfection and happiness – seen as synonymous – could be achieved through philosophical contemplation. All people return to the One, from which they emanated.

In the midst of all this, why the need to introduce a new practice of Japanese Zen? And that too in India? Even though India has the ancient wisdom of a structured science of sciences called epistemology (in the language of Western philosophy), through the various schools of thought or branches of Indian philosophy, somewhere down the line, we have lost the art of practice and the art of living. Our lives have become filled with the need to enhance the material quality of living. We strive to fulfil that, and in the bargain we get burned out. We are no longer in touch with any kind of wisdom frame of reference, even though we take pride in the spiritual wisdom of our ancestors. In the present-day world of science and technology, we have access to the entire globe within the palms of our hands. It feels outdated and pointless to hold on to geographical

calligraphy, Haikus and tea ceremonies. Non-attachment – Letting go and not grasping to anything. Understanding the impermanence of things. Breaking free from self-generated suffering of greed, hate and delusion. Silence – Silence plays a significant part in Zen practice. Allows for deeper introspection and direct experience of Reality beyond words and concepts. Rituals – Practice of bowing, chanting, choreographed ceremonies, which are performed mindfully. Giving the experience of expressing respect and devotion. Concern and care for the other. Deeper understanding of the true nature of existence.

How can Japanese Zen be fitted to suit Indian culture? The founder of Zen is Bodhidharma, who hails from Kāñcīpuram in south India. He carried with him the knowledge of the Dravidian martial art of that era called Koothuvarrisai and medical knowledge of Siddha Vaidyam. Basically, he taught the power of the mind. Which led to the founding of various martial art schools in China and Japan, like the famous Shaolin Temple in China, and other monasteries in both countries and other Asian countries that practise his teachings of silence. In today's Western world and the world of technology, Zen practice seems to blend well without conflict. In terms of a complete circle of Zen returning to India, it seems like the right time and clime, having stood the test of time, passing through various cultures of the world, Zen seems to be harmonizing as a "One World Practice", as it does not hold any dogmas or give anything to hold on to. For the religious or the non-religious, atheistic or theistic, for the sharp or the dull, commoner or scholar, lay or monastic Vedānta, Zen practice is a practice of Oneness. A kind of Dialectical Behaviour Therapy. Which is a practice of doing things together rather than talking about things together. Vedānta gives the theoretical understanding of the science of Oneness and about Wisdom beyond wisdom. Zen is the practice aspect of the same understanding, which is studied and applied through the Buddhist teachings which originated in India. It fulfils the need to get back in touch with our own body and mind, returning to the simplicity of ordinary mind, being comfortable to be the person that we are, familiar culture, in a practical and doable way, in the most universal or secular way, without being lost in blind faith or getting stuck in fundamentalist thinking, while being all-inclusive rather than being confined in boxes with labels. Zen in today's world has found ways to support engaged practice in all walks of life.

Narayana Guru, the great sage of the modern times, brought together the teachings of Vedānta and the compassion of the Buddha, which Nataraja Guru integrated into one world of unitive understanding of the Science of the Absolute, that Guru Nitya shared beyond boundaries as engaged practice of a Symphony of Values, which Guru Muni taught as a structured study of life and living.

May all beings be nourished. May all beings find ways and means to connect and align with the *dharma*. May all beings be supported to transcend and find harmonious ways to practise in the midst of *saṁsāra*.

21

Essence of Narayana Guru's Advaita Darśana

Gita Gayatri

THE word *advaita* means non-dual. How can *advaita* be different and unique to a person when it speakns of Truth as One? It is said that every person is a product of the times. Great Masters are like mirrors that reflect the need and spirit of the times. When austerity and penance became too severe and exaggerated, there appeared Siddhārtha Gautama, a kṣatriya, who rejected the dominant Vedic authority and became the Buddha and taught a unique philosophy that addressed the need of his time. When the priestly class controlled the whole social structure, and religious sentiments took an ugly turn, there came Śaṅkarācārya, who was tirelessly put down exponents of ritualism. He was no less ardent in speaking out against the flaws in Buddhism that undermined the ancient teachings of the *ṛṣis*. He intended to revive the wisdom of the Upaniṣadic *ṛṣi*s. He did not cultivate a philosophy per se, rather his main scope was to defend the ancient wisdom against many adversaries and he came up with rational arguments and explanations that stood by the scriptures. Scholars, however, have pointed out that he was influenced by Buddhism and that he even borrowed many elements from them.

When the scale on the Absolute side got weighed down by misunderstandings and misinterpretations by the Māyāvādins, there came Narayana Guru who straightened out the flaws that were projected and highlighted by the followers of Śaṅkara in their effort to make commentaries on ancient texts. When power and might became the right and man became a burden upon man himself and atrocities were committed in the name of caste, creed and religion, Narayana Guru simply applied his unique wisdom, not through force, but with skilful tact and compassion that touched the spirit of the times and brought about many changes to hardened ideas in many dimensions and aspects of life. In a structural schematic representation, wisdom is on the one end of the pole and the ugly face of distortion on the other end of the pole. Narayana Guru became the catalyst that finds a neutral balance in the centre that helped to enhance and nourish the spirit of the times on the vertical scale. Today in this age of technology and science, Narayana Guru's life and teachings have become a key factor that can be relevant and instrumental to hold together and support wisdom and action from a balanced neutral perspective leading to right understanding on a universal level. Verse 88 of *Ātmopadeśa Śatakam* says:

> Everything is real in itself; one who grasps the basic truth will understand this as one; if not known introspectively, *māyā*'s great enmity certainly creates much confusion.

One of the norms that Narayana Guru emphasized was *dharma*:

dharma eva param daivam, dharma eva mahā dhanam, dharma sarvatra vijayī ।
Dharma is the God Supreme, *dharma* is the greatest wealth, *dharma* becomes victorious ever. May it make humans blessed.

These verses Guru wrote as a benediction in Sanskrit for a magazine called *Dharmam*. The basis of Guru's teachings is rooted in the Upaniṣads. He accepts the basic doctrines of Gauḍapāda and Śaṅkara. We also find his affinity to the Buddha's teachings in his *Darśana Mālā* where the last chapter is titled Nirvāṇa Darśana rather than Mokṣa Darśana.

The basic and fundamental eternal questions are "Who am I? (*ko aham*)" and "What is this world that I experience? (*kuto jagat*)". This basic quest has been viewed from different dimensions through all schools of thought both Eastern and Western, through religions of the world, through science. What is it that brings us here again and again? There is a sense of value that connects us. The sorrows and the joys that we face in life when we touch the highs and lows give us a feeling of feverishness, a kind of *jvara* as the *ṛṣis* say. Vyāsa puts it through the words of Kṛṣṇa to Arjuna in verse 30, chapter 3, of *Gītā*:

mayi sarvāṇi karmāṇi saṁnyasyādhyātma-cetasā ।
nirāśīr nirmamo bhūtvā yudhyasva vigata-jvaraḥ ।।

Renouncing in Me all works, coming to be without expectations or possessiveness, with a full awareness about the Self, do fight with fever gone).

What is this *jvara*? Is it a self inflicted *jvara*? Our life is influenced by three factors. How do we experience this world, what are the tools through which we relate with the outer and inner world? There is something irrefutable about sensory experience. One or another stimulus is always trying to catch our attention. We are constantly in dialogue with the external world – *sat* (existence), what really exists – the reality. Second, there is a factor of "I know". Within our head there is a very restless machine which goes on reasoning. Constantly looking for meaning, for value that relates with all things and ideas – *cit* functioning through the inner organs (*antaḥkaraṇa) – manas, citta, buddhi, ahaṁkāra*. Besides observing and reasoning, there is a demand of the heart. There is a thirst which is neither of the senses or of the mind. There is a central value with which our sense of awareness must join. This demand of the heart is the third factor – *ānanda*. The first turns us outward, the second turns us inward, and the third relates and integrates the inner and the outer. This affectivity "I am affected" is born of an inbuilt value. In the *Daiva Daśakam* Narayana Guru defines God as *ne satyam, jñānam ānandam* (Truth, knowledge and value). When there is an imbalance in the harmony of these factors, we experience a sense of disease. Āyurveda's definition of good health is

> *sama doṣa, sama dhātu, sama agniśca, mala kriyāḥ, prasanna indriya, ātma, manas svāsthya iti abhidhīyate*
>
> Balance of energy, balance of wind, bile, phlem, balance of chemical and electric fires and impulses, balance in excretion of body wastes, cheerful senses, spirit and mind this brings lightness and good health.

As a defence mechanism, the body naturally has inbuilt fear, anger, ego, joy, etc. But when it becomes pathological and hinders our smooth functioning in our day-to-day life, we experience suffering of body, mind and depletion of spirit. We live in a consumer society where an active campaign to lure the senses through desire is going on all the time. Pain and pleasure are intimately connected with sensory agitations. In verse 8 of *Ātmopadeśa Śatakam*, Guru cautions us:

> *olimutalām pazam añcumuntu nārum, nalikayil eri nayena māriyātum* ।
> *kilikaleyañcumariññu kīzmarikum velivuruventiyakam viḷannitenam* ।।

> Enjoying the five fruits, such as beauty, mounted on a foul smelling gun and evasively flying back and forth, are five birds; having brought them down, through an inversion, that radiant inner awareness should fill ones entire being.

Zen training which is also the basic training of the Japanese and Chinese martial art, from which was born the famous Shaolin Temple, focuses on this radiant inner awareness to bring out the power of the mind more than physical strength. Our thirst to integrate the inner and the outer brings us time and again to listen to the nourishing and guiding words of Guru.

Here again on the value scale, Guru leads us step by step to right understanding through different analogies. Guru clearly gives practical guidelines by emphasizing creative action, right understanding through contemplation of scriptures and sense of value. So far we have seen the historical background which is the very ground where Guru's wisdom was applied and beneficially engaged. Thereafter on an individual plane we saw where we stand in relation to this in terms of our need on the other end of the horizontal plane. Integrating both into an applicable unitive and universal science is the essence of Guru's Advaitic *darśana* and how he unitively holds the very art of life in a wholesome holistic and practical way grounded in reality.

Why do we consider *brahma-vidyā* as a science – the science of all sciences? Why is this self-study based on science? How to distinguish science from non-science? Indian philosophy clearly distinguishes two different streams of thought in the ocean of wisdom, as *śruti* and *smṛti*.

While *śruti* comes from the source of knowledge or wisdom, *smṛti* comes from memory tainted by the conditioned state of mind. The world affects an individual in many ways, causing one to feel pain and pleasure. One is bound to this world, and hence seeks liberation. In all the Upaniṣads and wisdom texts like the *Bhagavadgītā* and *Yoga-Vāsiṣṭha*, there is always a sceptic, someone who seeks to know. Someone who is tormented and up against the wall. Not just a lukewarm curiosity for armchair philosophy. Someone with a question in mind, someone with a fire within. In Zen training, they say practice like your head is on fire. The *Bhagavadgītā* clearly says: do not waste time talking wisdom to people who do not need it. There is a qualification and a preparation. Devotion, discipline, sense of surrender are some qualities that are required. These prerequisites are seen in the opening verse of Narayana Guru's *Brahmavidyā Pañcakam*:

nityānitya vivekato hi nitarām, nirvedam āpadya sad,
vidvān atra śamādi ṣaṭka lasitaḥ,
syān mukti kāmobhuvi,
paścād brahmavid uttamaṁ praṇati
sevādhyaiḥ prasannaṁ guruṁ,
pṛcchet ko aham idaṁ, kuto jagad iti
swaamin vadatvam prabho.

Even through the discrimination of the lasting from the transient,
Attaining well unto detachment, the well-instructed one,
Duly well adorned with the six initial conditions known,
Such as calmness, control, and so on,

And thus keenly desirous of liberation here on earth;
He then greets with prostrations
A *Brahman*-knower superior,
Pleased and favourable by anterior attentions and service;
Thereafter should he ask of such a Guru:
"Oh Master! this "I" here, what is it?
Whence this world phenomenal?
Oh teach me this! great One".

To such a seeker, who seeks with a deep thirst to know, wisdom is imparted. There is a structure and a scheme. Science is open and universal. There are clear notions, as science seeks certitude. Third, it is specific, yet that specific knowledge can be applied to a general law that can unify all streams of knowledge. Science can stand on its own merit and does not require faith; each individual can verify it for themselves. Science is not private, the norm by which it can be measured is the same for all. In the development of modern science, today the physicist and the philosopher stand on the same platform. The believer and the sceptic are today on the same ground. On 6 April 1922, Albert Einstein and Henry Bergson publicly debated the nature of Time. This explosive debate transformed our understanding and reverberated across the twentieth century. It was a turning point when science overtook philosophy. Bergson's notions of elan vital and pure duration challenged both Charles Darwin's evolution and Einstein's theory of Time. Today, there is a revival in Bergson's ideas, especially after the findings of quantum mechanics which disagrees with the theory of relativity.

Nataraja Guru was perceptive enough to recognize the importance of Bergson at a time when Einstein's ideas were dominating the arena of science and Bergson was forgotten. Here is where Narayana Guru's grounded holistic teaching stands alive and relevant in this age of science and technology. The Upaniṣadic *ṛṣi* says: "By the taste of water, you can know all water, there is no need to empty the seven seas".

Narayana Guru points out deep science through simple analogies. Verse 10 of *Ātmopadeśa Śatakam* says:

> *iruḷil irippavanāru?colka nī yennoruvan urappatu keṭṭu tānumevam* ।
> *arivatināy avanoṭu nīyumār ennaruḷumitin prativākyam ekamākum* ।।

> Tell me who are you sitting in the dark? One asks. On hearing this, the other, in order to know, asks in return, who are you? The answer to both these questions is the same.

Most recently, Bernardo Kastrup, a physicist from The Netherlands, in his research for brain health came up with some ideas at a conference which are similar to Narayana Guru's profound wisdom in *Aṟivu*. He says:

> The body is an image of consciousness. Just that we know that we don't know it. If you subconsciously change your subconscious psychic life, physical state remains healthy.

He talks of the cosmic nervous system in "A Small Theory of Everything". Narayana Guru wrote his composition *Aṟivu* more than 100 years ago. In *Ātmopadeśa Śatakam,* verse 74, Guru states:

> *poṭiyoru bhūvil asankhyam appotikkul,*
> *ppeṭumoru bhūvitinilla bhinnabhūvam;*
> *jaḍamammarunadhu pole cittilum*
> *cittuṭalilum iṅṅ itināl itorkkil ekam*

> Countless are the particles of dust, that constitute the earth;

and the earth that exists in and through the particles
has no different being either.
Likewise, matter merges in the being of mind, and
mind in the material bodies. The two thus become revealed to be
One in essence when contemplatively perceived.

Guru Nitya poetically puts it this way:

> The mystic lives in the nuptial chamber of his divine Beloved and in this union the duality of the speck of dust and the universal ground is abolished.

Thus, we see *sat* as the cosmological, *cit* as the psychological and *ānanda* as the metaphysical realm of knowledge as represented by the scientists.

What is the position of God in Narayana Guru's Advaitic *darśana*? With this given body, to find its balance in this world we live in, fine-tuning each moment through the journey of life, we also need a sense of devotion and surrender. In the *homa mantram* Guru refers to *śreyas* and *preyas* – *śreyaś ca preyaś ca prayaccha svāhā* (please confer on us the higher values as well as the worldly values). Science is not comfortable with God. Fair enough! Science cannot be science with its continuous inquiry and search, if it all ends with a God. When someone asked Guru Nitya what is God, in one word he said "ātmāsphere". Narayana Guru's philosophy would bring us to a point where God alone exists and this world that we think exists separately is only a mistake in our understanding and attitude. Verse 4 of *Daiva Daśakam* says:

āẓiyum tirayum kāṭṭum,
āẓavum pole ñaṅṅaḷum,

māyayum nin mahimayum
nīyum ennuḷḷil ākaṇam.

As ocean and waves, wind and depth;
let us grasp inwardly the scheme of ourselves,
of māyā, Your glory and even You yourself.

In the physical world that which is gross, when analysed, leads to a causal factor. Salt, for example, gets analysed into sodium chloride, each of which can be reduced to molecules, and in turn to atoms, to energy, and finally waves. Quantum mechanics is now able to bring everything into wave form. Yet we do not live as waves. When we kick a stone hard, we get hurt and it starts to bleed. Guru integrates by equating the ocean with the myriad of creation. The wave is equated with *māyā* (our illusory perception). The wind that is directly responsible for producing the waves is equated with God's power to generate. There is no difference between the relationship of the wave to the ocean and the relationship of an individual to creation as a whole.

Ancient religions conceived God as Father or Mother. In Buddhism and Islam, God is not directly addressed. In Mahāyāna Buddhism and Sufism we see a tendency to resort to such images of intimate relationship as the Beloved or God's immanence seen in all creatures that coexists with man. These concepts are convincing and close to the heart so that one can turn towards the Divine with natural ease.

nīyallo sr̥ṣṭiyum, sraṣṭāvāyatum
sr̥ṣṭijālavam,
nīyallo daivame,
sr̥ṣṭikkulla sāmagriyāyatum.

> Are you not creation, the Creator too, as also the magical variety of created things? Is it not You again, Oh God! who is the very stuff of which all creation is made?

Last is his mystical yet humane composition *Piṇḍa Nandi* where Narayana Guru's devotion and sense of wonder merge into complete surrender to that eternal silence. Narayana Guru expresses his helpless state as a foetus in the womb of his mother. In verse 6, he says:

> *rethassu thane ithu rektham odum kalarnnu, nādam tiranduruvathāy naduvil kidanneyn, mathavum illa avide annu pithavum illen tātten vallarthiavan ān ivan innu śambho*!

> Semen indeed it was, mixed with ovum-blood,
> Along with pure sound attaining maturity,
> That assumed my form,
> That lay at a mid-region,
> No mother was there,
> And no father too.
> So raised by you alone, my sole parent,
> Is all that I am, here today.

Many are the mystical compositions of Narayana Guru. He brings Mother and Father together and surrenders his sense of union (*yoga*) to that higher power. In the last two verses of *Janani Navaratna Mañjarī*, Guru says:

> *sat aayi ninupari, cit āyi raṇḍum oru muthāyi monum arriyum, hrit āyi ninadhinnu vitthāyi vinnodu maruthāyi dr̥ṣṭi muthalāy.*
> *kothāyidum viśaya visthśramyennmathin attavumaayi vilasum.*
> *siddhaanubhuthiyillumyethaadhe yaamathimahattayidum jananni nee.*

> *bhuvādhibhutam adhināvasam illa, verum ābhāsamām idh arrivin*
> *nābhavishesham idhināvasamingullagil āvadhitam bhavathiyāl,*
> *nāvadhi tanvishayidhāvasamattu bhavadhavasam āke vilassum,*
> *dhyovān athinde mahimāv ārarrinju jananni!*
> *vazhttuvānum arudhe.*

In the beginning of the verse, Guru says:

> You do endure as *sat*, and *cit* also above it, and both together forming *ānanda*. And also as the mind that is aware of the three. You remain even as the causal seed of all these.

In the end of the verse, he says:

> Who is there to know the greatness Oh Mother! I find myself helpless in praising it, alas!

We have now seen the essence of Guru's philosophy from the two opposite poles of science and theology. In verse 100 of *Ātmopadeśa Śatakam*, Guru's philosophical wisdom cancels both realms to merge into the completeness of silence:

> *atumitumalla sadartthamallaham saccid*
> *amṛtamennu teḷiññu dhīranāyi, sadasaditi*
> *pratipattiyattu sattomiti*
> *mrduvāy mrduvāy amarnniṭeṇam.*

> Neither that, nor this, nor the meaning of existence am I,
> but existence, consciousness, joy immortal;
> thus attaining clarity, emboldened,
> discarding attachment to being and non-being,
> one should gently, gently, merge in *sat auṁ*.

Having said all this, and having gone over several dimensions of just an essence of Guru's teachings and wisdom, we come to a point where we see that "'All' is the starting point, the first postulate is 'All – ' the ground of all potentials". Then we see that

1. I am in it, all experiences, nothing is excluded.
2. I see, I feel objectivity and subjectivity.
3. I share with fellow beings – interconnection.
4. Many methods of measurement or norms can be applied.

Certitude comes when we have a norm with which to measure.

Certitude is necessary to hold on to knowledge with conviction. Fact and fiction lie interlaced, which need to be separated and distinguished in order to have right understanding. Without testimony, it is only an opinion.

Besides observing and reasoning, there is a demand of the heart, we feel the truth in our depth intuitively from within. This is the noumenal, in which the standard of measurement is neither of the senses nor the mind.

There is a central universal value with which our awareness connects. Senses turns us outward, reason turns us inward, the third integrates inner and outer (*yoga*). There is an orderly arrangement of *dharma*, reason wants to be free of emotional entanglements, truth is without compromise, there is a universality of Truth. The Sāṁkhya *darśana* uses these terms *puruṣa* and *prakṛti*. Who is this *puruṣa*? It is not the gender *puruṣa* as we understand literally. One who traverses through the three *puras*: *jāgrat*, *svapna* and *suṣupti* (wakeful, dream and deep sleep) smoothly is a *puruṣa*. Our job is to find a right balance in these three states of consciousness and the outcome of that will fine-tune us to connect with this world that we live in. *Prakṛti* is defined as *prakarṣena karoti iti prakṛti* (that which is properly engaged in activities). So, this given body has to find its balance physically, mentally and spiritually to live in harmony. "In the underuse, misuse and overuse of Time, sense objects and activity lie the roots of ill health, appropriate use thereof safeguards good health or *svasthavṛtta*", says Vāgbhaṭa, according to the science of Āyurveda.

Verse 72 of *Ātmopadeśa Śatakam* says:

kriyaorukūruitavidya; kevalam cinmayi marukūru itu vidya; māyāyle niyatamitiṅṅane nilkkilum piriññadvaya para bhāvana turyyamekiṭunnu.

Now there is action, which is ignorance, and again there is pure consciousness, which is knowledge;
although these two are thus ordained by māyā to stand divided,
a non-dual vision of the Absolute brings about *turīya*.

Guru thus points out three areas to look into:

1. The transactional world of dualities,
2. the contemplative realm where names, forms and events can be reduced to a single universal, and
3. the neutral ground where the transactional and the transcendental aspects are within the same reality.

Auṁ.

22

Living with Kokua, By Any Other Name

Peter Moras

On a recent stay on the Hawaiian Island of Oahu, I got to sink into the daily rhythm of the friendly residents of this ancient place of sea, salt, wind, rain, fertile plains, extinct volcanos, reefs, beaches, jungle and almost daily rainbows. One day, my friend and I visited the Bishop Museum in Honolulu. If you have not been there, it is a gorgeous, treasure house, filled with artefacts and information on the cultures of Oceania. I happened upon a special book entitled, *Pono: The Way of Living* by Malcolm Naea Chun. This book explores the basic values that underlie traditional, Hawaiian communities. Reading it got me to thinking about the rich meanings of the word *pono*, basically "right living", and how similar it is to core values cherished in cultures in other parts of the world, but which go by different names. It was interesting to read:

> There are no words in Hawaiian for "values", "morals", or "ethics". This lack is not uncommon in traditional cultures where moral or ethical practices were an integral part of everyday life. – Chun 2006: 1

A subsidiary value of *pono* is *kokua* which I would like to focus on for its everyday practicality and universality.

How are *pono* and *kokua* related? *Pono* is inspired by a profound sense of belonging with nature, interdependence with living beings and respect for *mana*. *Mana* is the spiritual life-force that animates and operates in, as, and behind what is visible. *Kokua* is the helpful spirit in a person who gives of itself out of the kindness of its heart. Lending a hand to a situation, person or place in need, without thought of reward or benefit, is *kokua*. *Kokua* can take many forms: informal, spontaneous and brief, or organized, broad-based and long-term.

Actions taken in this spirit of kindness which the Hawaiians call *kokua* is the same, selfless action for the sake of the action itself that the *Bhagavadgītā* recommends in its fifth chapter, Karma-Saṁnyāsa-Yoga:

> The one of unitive discipline, discarding benefit-motive, attains to ultimate peace; the one of non-unitive discipline, being desire-motivated, attached to results, is bound. – Yati 1993: 133-34

My teachers, while encouraging me to see and identify with the formless Absolute, in and behind the world's names and forms, stressed the importance of keeping my feet on the ground, loving and serving humanity. Lord Buddha showed a way to liberate ourselves from suffering, live in peace and find compassion in the stillness of our hearts, in the middle of the natural activities of daily life. Christ's example taught the importance of love, forgiveness and sharing. He urged us to remember that even the smallest actions taken on behalf of the most vulnerable and forgotten are done to the Divine in Him, that is the same Divine in us. Nataraja Guru of the Narayana Gurukula spoke of "unlimited liability" and "world citizenship" as a counterpart of Self-realization. Living *pono* and *kokua*, in this same vein, brings balance, integrity and well-being to relationships (Aloha Shirt Shop 2023: 6) and spreads peace, prosperity and security to communities. Imagine happy people who show recognition and respect for big and little things, like the fungal networks in soil, the tidy maintenance of public bathrooms and the purity and plenty of our water, air and food.

The positive energy of *kokua* is self-invigorating and self-regulating. Acts of service imply justice and bring happiness and satisfaction to the helper, while bringing benefit to the recipient (Aloha Shirt Shop 2023). It can be tutoring a child to read or an immigrant to learn a new language. It can be doing voter registration, mentoring a young person with fewer advantages than yourself, sharing a plate of cupcakes with your neighbour or reminding an elderly friend that they can always call on you for help. Think of the needs in your community or region. What do you feel like doing? It might be doing invasive plant removal, "fire smart" landscaping, community garden creation, urban tree-planting, fuel-load reduction in wildland–urban interfaces or habitat restoration projects in partnership with schools, civic agencies and non-profit groups. Want to see your streams and wetlands come to life again? Go for it! Make it happen! There is no end to the forms which *kokua* can take. Our sense of capacity and potential in our "being" increases through our "doing". The more we do, learning as we go, the more we feel we can do. Confidence, knowledge, skill and efficacy in problem-solving increase with baby steps.

As humans evolve, societies evolve. It is before us to mirror the goodness and wisdom implicit in Nature. We can receive her explicit message of beauty and the way she proliferates with astounding creativity, while conserving and ingeniously recycling materials and energies. We can notice how sunlight creates and regulates life on earth without asking for anything in return. We can be grateful for the quiet nobility of trees in forests, suburbs and cities that take in carbon dioxide and give out the oxygen we need without sending us a bill or putting their branches out for payment. They do that selflessly for the benefit of Gaia, the super-Organism. The ocean, mother of the sky we suck on, cycles water without which there are no life-giving clouds, rain, glaciers, rivers, aquifers or future. And no rainbows. The ocean is not anxious for commendations or recommendations. The ocean just is. It is true to itself. From my eyes, the ocean is an immense, impersonal, manifestation of the Supreme Absolute, the Compassionate Teacher in liquid form, watery sound and bathyal silence.

We can *kokua* without fanfare. It can reflect who and what we are, and how we want to be with each other. Like the sun. Like the trees. Like the ocean. Wherever we live. Whatever language we speak.

Thank you, dear Hawaii.

Here is a chant to *kokua*, from love in the *aloha*, the sacredness of pure Being:

- *AUṀ* "we wipe up the spill we notice in the refrigerator the first time we notice it", *namaḥ*.
- *AUṀ* "when we notice litter on the sidewalk or beach as we approach, we do the bend-over-and-pick-it-up *āsana*", *namaḥ*.
- *AUṀ* "we learn first aid and CPR, keep it updated, and offer it to persons in need", *namaḥ*.

- *AUṀ* "we make sure our children learn to swim early in life to be competent and safe to enjoy the water, and be capable lifesavers", *namaḥ.*
- *AUṀ* "when moving in slow or in stopped traffic, we allow drivers who are stuck waiting to enter or cross the flow of traffic from a side road, to do so safely", *namaḥ.*
- *AUṀ* "we do business with people who treat their employees well, and who show respect to people and places involved in their supply chain", *namaḥ.*
- *AUṀ* "we greet passersby with a smile, a warm hello, a *namaste* or silent prayer of well wishes to our Self in the form of the other", *namaḥ.*
- *AUṀ* "we listen to trees and watch breezes", *namaḥ.*
- *AUṀ* "we tell bushes, grasses and flowers we pass how beautiful they are", *namaḥ.*
- *AUṀ* "we scratch and pet our cats and dogs", *namaḥ.*
- *AUṀ* "we open our eyes underwater, wonder of wonders", *namaḥ.*
- *AUṀ* "wherever we are, we leave it more beautiful than when we found it", *namaḥ.*

AUṀ, AUṀ, AUṀ.

The state motto of Hawaii is "Ua mau ke ea o ka 'aina i ka pono".

"The life of the land is perpetuated in its righteousness."

– Chun 2006: 8

References

Chun, M.N. Pono, 2006, *The Way of Living*, Hawai'i: Curriculum Research and Development Group, University of Hawaii.

Yati, Nitya Chaitanya, 1993, *The Bhagavad Gita: A Sublime Hymn of Yoga Composed by the Ancient Seer Vyasa*, 2nd edn, New Delhi: D.K. Printworld (3rd revised edn, 2022).

Aloha Shirt Shop, 2023, "Pono Meaning", retrieved from http://www.alohashirtshop.com/blogs/hawaiian-lifestyle.

23

Vision 4023

The Future of Narayana Gurukula

Bijoyees O.B.

Introduction

CONTEMPLATING the future of the Narayana Gurukula, it becomes apparent that focusing solely on the physical structures of Varkala Gurukula or other Gurukula locations worldwide would be short-sighted. The relationship between teacher and student has been crucial in pursuing knowledge and developing philosophical traditions. However, the physical locations of these teachings have often been lost to time, leaving only their ideas to resonate in the collective consciousness.

Outline

In order to ensure that the essence of Narayana Gurukula endures for future generations, an innovative and visionary scientific story format has been devised, inspired by other philosophical institutions. Although the full story cannot be presented due to space constraints, it is hoped that a talented storyteller will transform it into a masterpiece for the Gurukula's upcoming 200th anniversary celebration.

Scientific

Some may be curious about how to approach the future of a philosophical school with scientific fiction creativity. The answer is simple: Guru Nataraja, the founder of the Gurukula, envisioned an Integrated Science of Sciences that seamlessly blends the Philosophy of Science and Scientific Philosophy, without any conflicts. His significant contribution to world knowledge was emphasizing the need for a "Normative Notion" to guide all sciences and philosophy, which he named the Absolute. Therefore, his Integrated Science is a profound exploration of the Science of the Absolute, which is the Science of all Sciences.

Future

To envision how the Gurukula might look in 4023, we need to consider what the world might be like at that time. There are two possible outcomes: a positive one where technology has solved many global issues and society is peaceful and prosperous. The other scenario is negative, with environmental disasters, social unrest and income inequality caused by technology. However, the future is unpredictable and can be influenced by factors such as politics, economics, and natural disasters. By contemplating different possibilities, we can prepare ourselves for whatever the future may hold.

Optimism

Looking towards the future with optimism, we can imagine a world in which technology continues to advance steadily, leading to new breakthroughs in renewable energy, medical science, space exploration and artificial intelligence by the year 4023. These advancements could have a significant impact on our world, potentially enabling us to explore and settle on new planets, discover new resources and technologies, and control population growth on earth. There could also be greater global cooperation, regardless of race, gender or social status, resulting in a more diverse and interconnected cultural landscape shaped by scientific, religious and philosophical beliefs.

Pessimism

It is important to remember that the universe undergoes cycles of growth and decay. While we may feel optimistic about technological progress, we should also consider the possibility of this cycle repeating itself in the future. As Einstein once said, he could not predict the weapons that would be used in the Third World War, but he knew that the Fourth World War would be fought with primitive tools like stones and sticks. It is crucial to contemplate this possibility before making any predictions or crafting stories. A catastrophic event, such as a global war, has the potential to destroy all energy grids, rendering current technologies obsolete and irrecoverable. A mechanical tool found as a fossil could be understood by anyone with basic knowledge, but unfortunately getting information from a quantum computer, whose coding is beyond common understanding, years after excavation, would be very challenging.

Relevance

No matter what lies ahead in the future, individuals will continue to be the most crucial element. Despite humanity's progress in technology and methodology, we still lack a comprehensive understanding of consciousness and everlasting happiness. Even if machines try to control humanity, people will always have their sense of existence and well-being. The Love of Wisdom has always been essential, and instead of questioning its importance, we should focus on spreading wisdom effectively. Nataraja Guru's "revaluation and restatement" method updates teachings to fit contemporary contexts. Thus, it is crucial to revaluate Narayana Guru's teachings to ensure the Gurukula's prosperity for many years to come. Rather than solely imagining the future, let us explore key considerations for keeping the Gurukula relevant in the following passages.

Absolutism

Naming a philosophical school of thought is an important aspect of its development and evolution. Nataraja Guru's philosophical school of thought, often associated with Vedānta, requires a specific name due to the wide range of ideas within Vedānta. "Absolutism" has been proposed as a unique and precise name that highlights the central concept of the "Absolute" in Guru's work, as demonstrated in his masterpiece *An Integrated Science of the Absolute*. This name accurately reflects the core idea of the philosophical school of thought, the Absolute, which distinguishes it from others.

Logo

Developing a symbol or logo to embody a philosophical concept is vital to its endurance over millennia. Nataraja Guru's emblem for Narayana Gurukula embodies fundamental human principles, including "life" and "love", using a fusion of artistic illustrations from both Eastern and Western cultures. The design incorporates the *AUM* syllable, the motto of unity, the ascending sun, triangles, stars and circles to symbolize boundless wisdom. This logo is a compelling and striking embodiment of Guru's philosophy and can play a vital role in disseminating his teachings.

Inclusiveness

Inclusiveness is a crucial element in establishing a philosophical school's future for millennia. The school must embrace diverse voices, perspectives, beliefs, values and practices while fostering a culture of openness and tolerance. This approach promotes relevance and applicability to people of diverse backgrounds and ensures a broader audience, increasing the likelihood of survival. When ethical principles, such as compassion and empathy, are incorporated into teachings, they are more likely to resonate across cultures and generations. Ultimately, inclusiveness leads to a better chance of being remembered and passed down, ensuring longevity for the philosophical school.

Youth

To ensure the longevity of a philosophical school, it is essential to influence the young minds of its time. Collaborating with secular organizations and integrating creative expression and entertainment can attract intellectually and aesthetically inclined young people to the organization. A relevant curriculum presented in their language is also crucial. Training young people to empower their juniors is important as well. If the school has a say in provincial governance, it could attract diverse groups of young people with different interests.

Expansion

For a philosophical institution to expand successfully, it needs a clear plan to avoid internal conflict and division. It can expand vertically and horizontally, mystically and academically, with various degrees between them, but must have a flexible framework based on core values to accommodate diverse interests and needs. Effective communication and collaboration between different sections can promote understanding and the exchange of ideas. By following these guidelines, the institution can grow while maintaining its integrity and coherence.

Preservation

To ensure the longevity of a philosophical school and its authentic foundational knowledge, it is important to preserve the works of the founder in their original form. To ensure the preservation of valuable primitive learning materials, it is essential to maintain physical copies in addition to digital versions. The original works of the founder should be safeguarded as they are prone to be overshadowed by commentaries. Sending the founder's complete works to libraries worldwide is crucial. Moreover, a sturdy room with long-lasting metal-printed versions should be established in various centres globally.

Globalization

For a philosophical institution to expand its influence globally, a strong online presence is essential. Translating texts into multiple languages, building relationships with other schools, and offering academic programmes, study groups, public lectures, debates, and art and literature creation can all help to engage a wider audience and foster understanding and collaboration.

Decentralization

While intellectual leadership is crucial for the success of a philosophical institution, it is equally important to create materials that emotionally connect with people. The institution should promote a philosophy that resonates with its followers and society as a whole, rather than solely focusing on the organization's interests. Decentralizing knowledge and making it accessible to the public are crucial steps, and the leadership should have a clear vision, communication skills and strategic planning to ensure that the philosophy belongs to the people.

Deinstitutionalization

To create a lasting institution, it is important to avoid turning it into a religion with unchanging ethics and strict rituals. While hierarchy and traditions provide stability, they can also lead to a dogmatic approach to philosophy. Charismatic leaders, with their own individualized philosophies, may steer the institution towards a more sectarian or moralistic direction. To preserve the institution as a philosophical centre, it is important to follow the founder's scientific methodology and ideas. This approach will ensure that Narayana Gurukula endures for generations to come without becoming a religious institution.

Conclusion

To conclude the science fiction narrative depicting the growth and life of the Gurukula over millennia, it is important to emphasize the process of revaluation. As an illustration of the significance of revaluation, a passage from *An Integrated Science of the Absolute* is presented:

> It is therefore in order to bring ancient wisdom more in line with modern thought that we have attempted a close study of one in terms of the other. They are meant to enrich each other and even if at present there might remain dark corners or vague speculations still calling for further clarifications or elucidation on more correct scientific lines, we can at least claim to have made an honest though humble effort to bring ancient wisdom more in line with modern thought. It is always possible for future research on the same lines to accomplish this more fully.
>
> – Nataraja Guru 2020, vol. 2: 478

Summary

To summarize, ancient wisdom does not aim to discover anything new or create new science. Instead, it emphasizes timeless wisdom that has been experienced by revered seers throughout history. It does not seek to control or dominate various domains of knowledge in society. In contrast, modern knowledge seeks constant new discoveries and relies on scientific methods and empirical evidence to assert its authority. However, the use of scientific methodology does not exclude future dependence on subjective perspectives that may replace the current objective outlook. A philosophical school should strive to balance objective and subjective aspects of reality using an Absolute "Norm" to unite them. The methodology of Nataraja Guru can be a valuable tool in achieving this goal in the proposed future functioning of the Gurukula, as long as the human brain structure remains the same.

Wishing everyone a great centenary year.

24

Twenty-first Century Reflections on the Narayana Gurukula

Compiled by: Nancy Saumya Yeilding

As Narayana Gurukula grows into its next 100 years, we can all be inspired by these twenty-first century reflections, representative of those shared by friends and students of many ages, countries and walks of life – poets, computer scientists, artists, social workers, business owners, filmmakers, therapists, teachers, journalists, medical professionals, and more – reminding us of the transforming power of loving wisdom:

> At the Narayana Gurukula I discovered the simplest of living yet the highest of thinking. The books, the teachers, the students, the mirror that reflects the infinite treasure residing deep within. Love.
>
> – Sanal Madhavan (UK)

> Narayana Gurukulam is a nurturing space where I feel safe, joyful and peaceful to further explore and grow amidst Guru's words and presence.
>
> – Chithra Ram (USA)

I was very fortunate to have known Guru Nitya at a young age. Nitya and Vijay were classmates at Union Christian College, Aluva. Guru Nitya really inspired our family with wisdom and wit. He was indeed the spiritual conduit for Narayana Guru, and I am trying to live a life following many of Guru's teachings. It has made remarkable changes in my personal life.

– Girija Pillay (Indiana)

A guide to the wandering soul to connect with the inner light.
A guide to seek the inner light through experiencing the universe.
A guide to experience I in others.
A guide to see the unified I in the collective individuated I.

With love and thanks.

– D. Gopica (Tamil Nadu)

I am appreciative of the wisdom passed along from the sages of Eastern Universal Truths. An evolution of songs and oral histories of Narayana Guru through his students. Always thankful for your particular sharings. The title alone, *Autobiography of an Absolutist*, is an expression worth pondering. An insight into an all-inclusive wisdom. A great reduction of a concept into its simplest form. In its wording one can sink into the silence from which the mystery arises. Love.

– Bob Lucas (Costa Rica)

Since my parents met under the tutelage of Guru Nitya Chaitanya Yati, my association with Narayana Gurukula has been the foundation upon which I was born. It has served as a touch-point for continued exploration of philosophy, psychology and spirituality. Lastly, it has provided me with an extended family circle.

– Lakshmi Sarah (Berkeley, CA)

Regular reflection on connection and meaning helps keep me stabilized and doing it with a group of friends is most beneficial. The three guiding words I carry with me daily are balance, perspective and interconnection.

– Jean Norrby (Sweden)

I have never met Guru Nitya yet he has shaped me in innumerous ways. I am fortunate to have the opportunity to study closely with a lot of his direct students – Nancy, Scott, Deb, Andy, Johnny, Moni, Peter, Bill, Jayamohan – and the list grows. I take this moment to express my gratitude to you all for accepting me into the family and nurturing me. This moment would not have been possible without Guru's blessings. I prostrate myself at his feet. Thanks.

– Prabhu Muruganantham (Oregon)

With Nancy as our instructor and guide, a small group of participants and I studied the 100 verses very closely and received an introductory education over five years to Indian philosophy. I was particularly moved by Nancy's dedication and absolute devotion to Guru Nitya and the love she shared with us through his teaching.

– Viktoria Vidali (Hawai'i)

It has been my life's single greatest blessing to have floated into the orbit of Guru Nitya around age 11 in 1981 via my maternal grandfather's association with Nataraja Guru and Narayana Gurukula. Through meetings and interactions in Bangalore (where my family lived) and Fernhill Gurukula, peppered over the next eighteen years till Guru Nitya's *samādhi* in 1999, I received his love, teachings and blessings, not realizing the extent to which they would transform and inform my life. It was only after Guru's *samādhi* (by which time I was in California), that I began to water and nurture the seeds he had planted in me. I embraced the larger Narayana Gurukula family and formed close relationships with many of Guru

Nitya's disciples in the US and India who continue to be family to me to this day. Nancy has been my primary mentor and guide these past twenty years, while also being a very dear friend. Through her remarkable efforts and patient guidance, I was able to participate in online studies of *Ātmopadeśa Śatakam*, the *Yogasūtra*, *Bṛhadāraṇyaka Upaniṣad* and now *Darśanamālā*. These contemplations and other spiritual practices have opened my heart, purified my mind, and brought beauty, love and meaning to my life in ways I could have never imagined or predicted. I will forever be grateful for the wisdom and grace of our *gurus*, and the love of my Narayana Gurukula family.

– Gayathri Narayanan (Palo Alto, CA)

It is not easy for a kid born into an ordinary family in Kerala to feel at home in a city like London but also be equally appreciative of the life at North Paravur, a small town in Kerala. I thank my stars and the ancestors for the initial introduction of the concept of the East West University through the books of Nataraja Guru and Nitya Guru, and reap the benefits of it later in life by attending the online classes of Scott Teitsworth and Nancy Yeilding which has helped me to stay grounded and level headed while dealing with the freedom of the Western culture and the restrictions of the Eastern culture! Namaste.

– Vizzy Sagar (UK)

Like many others, a speechless gratitude comes when thinking of the impact that the love and teachings of Narayana Gurukula has had on my life. Through online study courses with texts such as *Ātmopadeśa Śatakam*, the *Bṛhadāraṇyaka Upaniṣad* and *Darśanamālā*, and connections with direct students of Guru Nitya Chaitanya Yati, I have often felt like I have one foot in a storied past, and another in how the teachings bring me perpetually back to the present in how to live and apply them. I have had the great fortune of seeing how to live this loving wisdom through the example and guidance of my aunty Nancy. – Robert Yeilding (California)

I am fortunate enough to be part of *Ātmopaeśa Śatakam* group. Internalizing each verse enables awareness to nurture the observing Self so that my thoughts and behaviours become guideposts, a way to understand and release conditioned reactions to perceive things with better clarity, insight and calmness. Thanks.

– Pratheepa Shivakumar (Tamil Nadu)

"Knowledge is incomplete without a Guru." I have been exposed to the teachings of Narayana Guruji, who may not be with us physically but the reflections of various aspects of life have been explained so beautifully in various verses, that I find my spiritual journey very fulfilling and enjoyable. While reading, suddenly a door opens up to show you something different which you never thought about. It makes you think and realize as to why a particular thing happens in this universe. The format devised by Nancy and her painstaking efforts to amplify the thoughts and contents of the verse further, as also share reflections of other members in the group makes the entire experience more holistic. My salutations to the Guruji and Nancy!!

– Anil Garg (Maharashtra)

Guru Nitya was part of my life from early childhood and I loved him dearly. I feel blessed to have been born into the worldwide family of the Narayana Gurukula and connected to the "Knowledge which brilliantly shines" through the incredible study groups led by Nancy aunty. To know also, that the Gurukula is so much more than a set of buildings, that it is also located within my heart, mind and essence.

– Emma Walker (Australia)

A brief celebration of my life with Narayana Guru
Even though he was born years before I was in this lifetime,
it's marvellous to see how he has reached me.
Nothing is linear, it's all a chaotic squiggle.
All connected.

Deep gratitude – Shyla Rao (Tamil Nadu)

Following an urge from inside, I visited Narayana Gurukula during October 2013. I felt the right place for me to contemplate which led me to buy as many books that I can from the bookstall. The clarity of Guru's presence that I imperienced and experienced which I treasure guided me to learn (from Gurukulam books) towards achieving that goal of clarity and still continuing. – Dr Suresh Sreedharan (UK)

I grew up in a family where Narayana Guru was worshipped, Nataraja Guru was my father's Guru and Guru Nitya his friend. From the time I remember, celebrated every birthday of Nataraja Guru in Bangalore Somanahalli Gurukula. I met Guru Nitya when I was 8 years old and life changed. Guru was my protector and I surrendered to him. With his love and blessings, overcame everything. Vacations spent in Fernhill Gurukula was bliss. Met many disciples and became part of the big Gurukula family. After Guru's *samādhi*, I was struggling and the online study with Nancy aunty got me back – her words of wisdom, guidance and encouragement – to move forward gives me immense strength. Thankful now to Friday meditations. I attribute who I am today to the blessings of Guru. In gratitude,

– Girija Kalidas (UAE)

From the classes with Nancy, *Ātmopadeśa Śatakam*, *Living the Science of Harmonious Union* (*Yogasūtra*) and *Darśanamālā*, I have and am learning, every month, and moving towards greater peace and joy being in the moment. Wisdom and good actions, and the meaning, are clearly alive and breathing through these classes given with such dedication, commitment and love. The effect of it is creating positive ripples that stretch beyond the classes and its members, far wider, and moving each of us towards *brahma-vidyā* – each to their own way to self-realization. What a gift, and thank you, Nancy.

– Saila Anandan (UK)

Charles Erickson with his painting of Narayana Guru in honour of the Centenary of Narayana Gurukula. Photograph by Brenda Erickson

Narayana Gurukula Publications in English

Books by Nataraja Guru

- *The Word of the Guru*
- *Bhagavad Gītā* (Commentary)
- *Integrated Science of the Absolute* (2 vols)
- *Saundarya Laharī* (Commentary)
- *One Hundred Verses of Self-Instruction* (Commentary)
- *The Autobiography of an Absolutist*
- *Dialectics*
- *Unitive Philosophy* (consisting of shorter works: *Vedanta Revalued and Restated; The Philosophy of a Guru; The Search for a Norm in Western Thought*)
- *Experiencing One-World* (consisting of shorter works: *Memorandum on World Government; One-World Economics; One World Education; Religion and Ethics; Unified Science*)
- *Wisdom: The Absolute is Adorable*
- *An Anthology of the Poems of Narayana Guru*

Books by Guru Nitya Chaitanya Yati

- *Bhagavad Gītā* (Commentary)
- *Saundarya Laharī* (Commentary on *Ānandalaharī*)
- *That Alone* (commentary on *Ātmopadeśa Śatakam*)
- *The Psychology of Darśana Mālā*
- *Garland of Mystical Visions* (commentary on *Darśana Mālā)*
- *Living the Science of Harmonious Union* (commentary on *Patañjali Sūtras*)
- *The Bṛhadāraṇyaka Upaniṣad* (commentary in 3 volumes)
- *Neither This Nor That But Aum* (commentary on *Ātmopadeśa Śatakam)*
- *Narayana Guru* (a brief biographical sketch)
- *Narayana Guru* (published by the Indian Council of Philosophical Research)
- *Love and Blessings* (Autobiography)
- *Love and Devotion* (Commentary)
- *Meditations on the Self*
- *Meditations on the Way*
- *In the Stream of Consciousness*
- *Experiencing the Īśāvāsya Upaniṣad*
- *Īśa Upaniṣad* (commentary)
- *Nataraja Guru and I*
- *The Psychodynamics of Praṇava*

- *Praṇava: Harmonisation of the Vital Energies*
- *My Inner Profile* (poetry)
- *Haunting Echoes of Spring* (poems)
- *A Bouquet of Verses in Praise of the Supreme Mother*
- *Gita: A Managerial Science*
- *Guide to Hindu Iconography*
- *Psychology: An Eastern Perspective*
- *Beyond Cause and Effect*
- *God: Reality or Illusion*
- *Marxism and Humanist Monarchy*
- *Epistemology of Gnosis* (commentary on *Arivu*)
- *Vināyakāṣṭakam* (commentary)
- *Universal Prayer* (commentary on *Daiva Daśakam)*

Books by Guru Muni Narayana Prasad

INDEPENDENT WORKS

- *Karma and Reincarnation*
- *Basic Lessons on India's Wisdom*
- *Functional Democracy: A Failure in India*
- *Children's Narayana Guru*
- *The Philosophy of Narayana Guru*
- *Pure Philosophy Simplified for the Youth*

- *Christ the Guru*
- *Three Ācāryas and Narayana Guru*
- *Gurukula Dhyāna Mañjūṣā: A Bouquet of Meditations* (compilation)

COMMENTARIES

- *Life's Pilgrimage through the Gītā* (2005)
- *Vedānta Sūtras of Narayana Guru* (1997)
- *Vedānta upto Narayana Guru*
- *Īśāvāsya Upaniṣad* (2009)
- *Kena Upaniṣad* (1995)
- *Kaṭha Upaniṣad* (1998)
- *Praśna Upaniṣad* (1999)
- *Muṇḍaka Upaniṣad* (1998)
- *Māṇḍūkya Upaniṣad* (2009)
- *Taittirīya Upaniṣad* (1994)
- *Aitareya Upaniṣad* (2000)
- *Chāndogya Upaniṣad* (2006)
- *The Complete Works of Narayana Guru*
- *Garland of Visions* (commentary on *Darśana Mālā)*
- *Life and Lessons with a Guru* (memories)
- *Shorter Philosophical Poems of Narayana Guru* (2010)

Books by Swami John Spiers

- *Guru the Unknown*
- *What Shall I Read?*
- *Philosopher Saint* (collection of essays from *Values* magazine)
- *Pagan Europe* (unpublished)

Books by S. Omana

- *The Philosophy of Narayana Guru*
- *Vedanta: The Science of Consciousness*

Books by Scott Teitsworth

- *Krishna in the Sky with Diamonds*
- *The Path to the Guru*
- *Liberating Ourselves*
- *Coming Back to Ourselves*
- The following websites are also run by him:
  — Nitya: http://aranya.me/index.html
  — Scott: https://nitya-teachings.weebly.com/

Books by Nancy Yielding

- *Narayana Guru: A Life of Liberating Love*
- *What Narayana Guru Is Not*

Books by Peter Oppenheimer

- *Mirror by the Road: A Transforming Journey of Spirituality in Everyday Life*, Inner Wealth Press (1988)

- *Keys to Eden: A Search for Meaning and Celebration at Home, at Work and on the Road*, Inner Wealth Press (2004)
- *Windows, Mirrors and Doors: A Living Appreciation of Narayana Guru's One Hundred Verses of Self Instruction*, Narayana Gurukula (2019)
- *24 Truths a Second: A Treasury of World Cinema, 1987-2021* (2022)

Books by Swami Tyageeswaran

- *Maha Guru*

Books by Sadhu Gopidas

- *Of Love and Colours* (poems)

Books by Deborah Buchanan

- *Gestures in Silence: Who Narayana Guru Is*

Books by Vinaya Chaitanya

- *A Cry in the Wilderness: The Works of Narayana Guru* (tr. Vinaya Chaitanya) HarperCollins India (2022)

Contributors

Guru Muni Narayana Prasad is Guru and Head of Narayana Gurukula. He has authored more than 120 books. He has been honoured with Padmasree by the Government of India in 2024.

Swami Thyageeswaran is Regulating Secretary of Narayana Gurukula. He has authored many books.

Swami Tanmaya is a seeker of the Absolute who maintained a bipolar relation with Guru Nitya from 1985 to 1999. Continuing in the Gurukula now with a bipolar relation with Guru Muni Narayana Prasad, from whom he received *saṁnyāsa* on 1 January 2002.

Dr Sabu is Chairman, Centenary Celebrations, Narayana Gurukula.

Nancy Yeilding is founder and head of Island Gurukula Aranya — a universal contemplative center — in Washington state. She is a lifelong dedicated student and teacher of the world's rich heritage of unitive wisdom, particularly as exemplified and taught by Narayana Guru and his successors. From the centre, which also functions as the Western headquarters of the East-West University of Unitive Sciences, she reaches students around the world via Internet-based classes. Editor of dozens of books on spirituality, science and health, she also was the long-time publisher of the English *Gurukulam* magazine, a journal of philosophy and arts. Living amid a garden of herbs, flowers, vegetables and fruits, she nurtures and is nurtured by nature's abundance and beauty.

Scott Teitsworth is a lifelong student of Indian philosophy in the lineage of Narayana Guru, and an editor of English-language books written by Guru Nitya Chaitanya Yati and Nataraja Guru. The author of four published books of his own, he and his wife, the poet Deborah Buchanan, have a Portland, Oregon, *āśrama* where they have presented in-depth classes since 1978. Scott teaches online classes, and his extensive writings are available on his free website. With a unique name, he is easily searchable on the Internet.

Deborah Buchanan has been a life-long student of Guru Nitya Chaitanya Yati and the teachings of the Narayana Gurukula. For over forty years, she and Scott Teitsworth have hosted weekly classes at the Portland Gurukula. She has also been a teacher at a science museum, in English as a second language classes, creative writing workshops and on Asian art history. For many years she was a serious gardener with an herb business. For the last twenty years, poetry has been her way of seeing the world, and she has collaborated with the composer Lisa Ann Marsh, her poems are song lyrics. Her family and two daughters have been an essential delight of her life.

Peter Oppenheimer is a teacher, psychotherapist, photographer and author, who stumbled into the Narayana Gurukula in 1971, as the result of what he has come to call "a series of divine mistakes". Since then, the teachings of the Narayana Gurukula, and Guru Nitya Chaitanya Yati in particular, have become the lodestars of his life.

Swamini Brigitte was born in Belgium. Joined the Narayana Gurukula in **Ezhimala**, Kerala in 1969. Given *saṁnyāsa* in 1970 by Nataraja Guru.

Swami Vyasa Prasad is a seeker interested in science and philosophy under the aegis of Unitive Understanding.

Dr Omana is a disciple of Nataraja Guru, Guru Nitya Chaitanya Yati and Guru Muni Narayana Prasad. Her doctoral thesis was on "The Philosophy of Sree Narayana Guru", and post-doctoral research project focused on "Epistemology of the Guru". She has authored several books on Narayana Guru's life and philosophy, and contributed manifold papers covering various aspects of Guru's philosophy at different national and international seminars and publishing her papers in national and international journals. She has guided and produced several PhD theses. Since retiring as Professor of Philosophy at the University College, Trivandrum, she spends much time teaching and working upon Guru's philosophy.

Dr B. Sugeetha is former Director of International Centre for Sree Narayana Guru Studies. Authored eight books about Guru Darśana. Now a senate member of Sree Narayana Guru Open University.

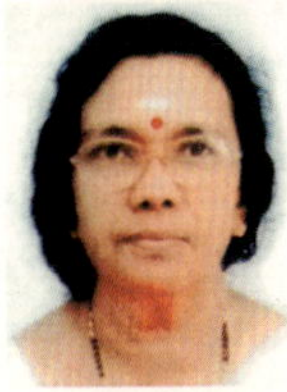

Dr Prabhavati Prasannakumar is a retired Professor of Political Science. Has authored many books.

Dr Shareena Banu is a poet and independent researcher on Narayana Guru.

Swami Mangalmurti is a spiritual seeker and teacher living in Varanasi, Uttar Pradesh. He respects all religions, specially Buddhism for its humanitarian and global vision. He has been awarded a doctorate from the University of Allahabad and has published six narrative poetical books. An epic in poetry on the life of the Buddha is yet to be published.

Deepanjali is a *gṛhastha śiṣya* of the Narayana Gurukula. Blessed by the love and care bestowed by Guru Nitya, Guru Muni and the subsequent Guru *paramparā*. Seeking to listen to inner Guru.

Gita Gayathri has a long time association with the Gurukula, from her age of 3 when she first met Nataraja Guru and Guru Nitya. She was a disciple of Guru Nitya Chaitanya Yati. She moved to California, to the San Francisco Zen Centre after Guru's *samādhi* and was ordained as a Zen priest. She was also given *saṁnyāsa* in the Daśanāmī lineage of Ādi-Śaṅkarācārya from Gangotrī Himalayas. She is presently researching Vedānta and Zen, through the medium of studying and teaching.

Peter Moras was seventeen when he met Guru Nitya, then "Swami Nitya", and was guided by him to become a teacher. Guru helped him see that God was not far away, but the everpresent, overseeing, eternal light of Being, also the dirt. While semi-retired from teaching, he enjoys helping with landscaping, composting and cleaning projects, travelling, and being in the kitchen!

Bijoyees has been affiliated with the Narayana Gurukula since 2009 and has been a full-time member since 2013. From 2014 to 2022, he served as the main aide to Guru Muni Narayana Prasad, the head of the Narayana Gurukula. He was the main associate in the ten-year-long translation of *An Integrated Science of the Absolute* into Malayalam by Guru Muni Narayana Prasad and was also a member of the editorial board for the 2020 edition of the original English version of the book. Currently, he is engaged in exclusive studies on Nataraja Guru's structuralism and dialectical methodology while pursuing the development of an extensive website dedicated to Nataraja Guru's work, utilizing large language models (LLM) to provide tailored answers to questions asked on the site.

Index